Basic Electrical Troubleshooting for Everyone

Darrel P. Kaiser

Published July 2007
Darrel P. Kaiser

Darrel Kaiser Books
www.DarrelKaiserBooks.com
email:Dar-Bet@att.net

First Printing

ISBN 978-0-6151-5684-2

Notice

This book is a complete course explaining the basic and logical process of troubleshooting and fault diagnosis. This book is generic, and does not address specific model electronic, component or appliance failures.

This instruction is not an "authority" to perform any repairs, or a recommendation to perform any repairs. This course is for educational purposes only.

Always refer to the manufacturers' current technical specifications, safety and repair manuals for the specific warnings and instructions applicable to the item(s) you are troubleshooting.

No warranty or representation, express or implied, with respect to accuracy, completeness, or usefulness of the information contained in this document, or use of any information, apparatus, method, or process disclosed in this document may infringe on privately owned rights.

No liability is assumed with respect to the use of, or for damages resulting from the use of, any information, apparatus, method or process disclosed in this course.

The Author

Darrel P. Kaiser has been professionally trouble-shooting electrical, electronic, and mechanical components and systems for the US Government for the last 37 years. During those years, he also trained with PFAFF in Germany and Bernina USA in the art of professional sewing machine repair, and continues repair and restoration even today.

He has also been researching the development of the Germanic peoples and his ancestors for over 10 years. While living for over two years in Germany, Darrel "walked the lanes" and did on-site research in the villages of his ancestors.

After all those years of troubleshooting and repair, he turned to teaching at a Government University and writing technical books. Out of his research came his first book on Germanic History and Genealogy, "*Origins and Ancestors Families Karle & Kaiser of the German-Russian Volga Colonies.*"

Darrel has also written and published numerous other books on German and Russian History, Politics, Religion, and Ancestry; a book on the Watercolor quilts of Betty Kaiser, a book on basic electrical troubleshooting, a book on sewing machine troubleshooting, two books on the SINGER 221 *Featherweight*, and two books on the STANDARD *Sewhandy* and GE *MODEL A* sewing machines. This book's final pages show all the titles.

For more on his research into German and Russian History and Genealogy, visit his website at
www.Volga-Germans.com

For more on his books on Troubleshooting, visit his website at
www.BasicTroubleshooting.com

For more on his books about Sewing Machines, visit his website at
www.SewingMachineTech.com

For more on his books about the STANDARD *Sewhandy* and GE *MODEL A* sewing machines, visit his website at
www.SewhandySewingMachines.com

For more information on all of his books, visit his website at
www.DarrelKaiserBooks.com

Preface

What does the title of this book mean?

It is the idea that we can approach any electrical or electronic, and mechanical fault using a basic logical or probability-based investigation to observe and correctly identify the significant indicators that will eventually lead us to the failure or failures.

This is no different from the Detective Books you read, or TV Shows you watch, where the hero uses a logical approach (while all those around him just run around willy-nilly) to identify the clues and catch the bad guy.

This book is a complete course in Basic Electrical Troubleshooting. Along with the written theory explaining my troubleshooting methods, there are over 80 diagrams and drawings, and 50 comprehension questions (with the answers) that will help you monitor how much you understand.

Table of Contents

Introduction to Troubleshooting

I have been troubleshooting electrical and mechanical equipment for over 45 years, and this book is the end product of all those successes and mistakes (or at least all those I still can remember at my age).

Most books on Electricity and Electronics will start with the very boring basics. Something like page after page on Current like this: "Composition of Matter – Controlling the behavior of electrons is what electronics is all about. Therefore, an understanding of the electron is vitally important to an understanding of electronic fundamentals." Or there may be many pages on Electricity something like this: "The word 'electric' is actually a Greek-derived word meaning AMBER. Amber is a semi-translucent (semitransparent) yellowish mineral..."

I am not suggesting that learning and understanding the electronic formulas is not important. I am saying that while all that detailed level of knowledge is necessary to correctly design electronics; it is not required to be able to logically and efficiently troubleshoot electrical circuits. I am not saying that the added knowledge would not be helpful; I am just saying that it is not a required prerequisite for success.

The chapters in this book cover the techniques that I found useful during my career troubleshooting everything from antique cars to cook tops to industrial electronics to communication systems to missile systems to laser systems to modern attack aircraft. I have learned that it does not really matter what the item is. What matters is your approach.

OK, I will agree that a system on an attack aircraft is more complex than that on an antique car, but the troubleshooting approach is the same.

Systems that appear to be complex only need to be broken down into smaller and easier to understand individual circuits. These individual circuits can then be broken up and isolated into even smaller sections. At this level, you will find it easier to grasp the circuit operation and testing procedures.

Troubleshooting is a skill that takes practice. I was taught and still do my troubleshooting from the positive to the negative. That is the method I use in this book. I hear you saying "But the electron flow is from the negative to the positive." That is true. It is just easier for me to do it my way... If you want to trace negative to positive, do it. We will both arrive at the same answer.

This book shows you how to do all of the above.

A Few Basics

There are some basic electrical concepts that you need to understand. You may know the following information. If you do, skip this section and move on.

<u>Current</u> is the flow of electricity. It is measured in Amperes or Amps. One amp of current represents a measured amount of electrical charge moving past a specific point in one second. Simply put, this is similar to gallons in plumbing. Current is divided into two types: AC or Alternating Current, and DC or Direct Current.

<u>AC</u> flows in one direction, reverses, and then flows in the opposite direction. In your home electrical circuits, this reversal happens 60 times a second or 60 cps (cycles per second). This is also known as 60 Hz.

European homes generally use 50 cps, and the two are not compatible. Military aircraft use 400 cps because the effective power output is greater with higher cps.

<u>DC</u> flows in one direction. Your car battery is a good example of this. It has positive and negative terminals that always are the same polarity.

Voltage is the difference in electrical charge between two points in a circuit. This is measured in volts as in 120 Volts. Simply put, voltage is similar to pressure in plumbing. It is the "push" in electricity.

Wattage is the amount of electrical power and is measured in watts. It can be computed by multiplying the voltage by the current, i.e. 120VAC x 15 Amps = 1,800 watts.

Resistance is a material's opposition or resisting to the flow of electrical current measured in ohms. The simplest example is a resistor, but all loads such as motors, transformers, and light filaments create a resistance to the electrical flow. Without this resistance, runaway current would quickly destroy the circuit.

A Series circuit supplies electric power to any number of devices or loads so that the same current passes through each device in completing the path to the supply return. In the example below, the light bulbs represent the devices or loads in series.

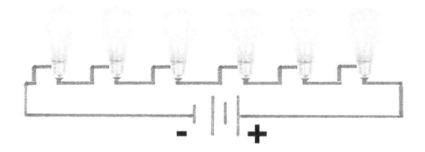

A Parallel circuit supplies electrical power to two points with all loads connecting between those two points as separate circuits. (Example below).

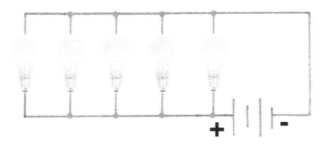

A Series-Parallel circuit has some of its components connected in Series, while others are connected in Parallel. (Example below).

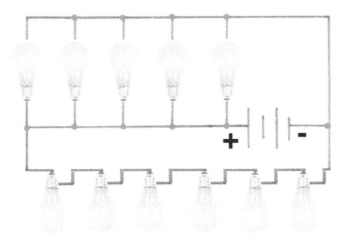

A Redundant circuit is a second path duplicating another circuit path. These are not common in normal consumer electronics, but are very common in military equipment. This design insures circuit operation and increases survivability when there is a risk of battlefield damage. Redundant circuits present special problems in troubleshooting, i.e. a damaged and shorted redundant circuit could prevent the original circuit from functioning properly, or vice-versa.

Load is the resistance in a circuit, i.e. a resistor, a motor, a light bulb, a transformer, etc. Without the Load resistance in the circuit between the Hot and the ground, you would have a Short with runaway current and circuit destruction.

An Open is a break or disconnect in a circuit. It could be a fault like a cut wire, or an open caused by a switch or circuit breaker. An Open causes the total circuit resistance to become infinite and reduces the total circuit current to nothing or zero.

A Short is an unplanned connection of the hot side to the ground side with no resistive load in the circuit.

Another way of saying this is that a Short occurs when two points with different voltage levels (i.e. hot and ground) are connected with no resistance or load between them. This occurrence reduces circuit resistance and allows for runaway current that will quickly destroy the circuit.

A Short causes more current and less resistance. It is this increase in Current that destroys the circuit.

The Troubleshooting Approach

Most people have a fear of electricity. Over the years, the public safety messages, our teachers, parents, and friends have pounded into our heads how dangerous electricity is, and how you will probably die if you get anywhere near it. I know people that are afraid to change their own light bulbs.

Some of the warnings are true... Electricity can kill, just as a car will kill if used improperly. Or just as something as simple as a knife, if used incorrectly, may cause you to cut yourself and quickly bleed to death. There are rules to follow for cars, knives, and for electricity. All of these items require you recognize and be aware of the dangers and risk, and their proper safe handling procedures.

On top of the reinforced fear factor, the sheer complexity of electrical and electronic circuits intimidates most people. And then there is the dollar cost of messing something up.

Just as there are safety rules to minimize risk of injury, there are troubleshooting rules to minimize the complexity and the risk of causing damage. There is also a wide variety of inexpensive test equipment that will help in our troubleshooting adventures.

I am not minimizing or downplaying the very real danger of working with electricity. It can easily KILL. Even if you follow <u>ALL</u> the common-sense rules, there always remains a risk of being shocked.

Nor do I want to downplay the costs that faulty troubleshooting can cause if you burn out an expensive component. Be cautious. Remember the wise old saying for carpenters that went "Measure twice, Cut once." For anyone working with electricity it might be "Measure twice, Think twice, and only then do something."

One of the abilities that you will need to master is the one that tells you when you are getting in too deep and over your head, and time to call in someone more experienced! Do not feel bad if you have to call someone in that is more experienced to solve the problem. I know that Ego, Pride, and everything else gets in the way, but life is way too short to worry about those minor issues.

Your goal should be to observe and learn every thing you can from the more experienced person. Make a learning experience out of every repair. The next time (and there usually is a next time) the problem shows up, you will already know how to approach and troubleshoot it.

Personal Safety

Little of the safety information in this section will be new to you, but it will not hurt to refresh your memory. It may even save your life, so here it is again.

Electricity will cause fatal burns or make vital organs in your body no longer work properly if you make a fatal mistake.

Most people have experienced a tingle when exposed to electricity. Many of you have experienced the discomfort caused by the group joker sliding his feet on the carpet to build up static, and then touching your ear. Obviously, it does not take much of a shock for us to take notice.

Generally, a current of five mA (milliamps) or less will cause a shock sensation, but will rarely cause any damage. Larger currents can cause muscle contraction. Currents as low as 100 mA for even a few seconds can be fatal depending on the voltage and how and where they pass through your body.

That ever-present 110/120 VAC home wall socket circuit is capable of 15 to 20 Amps (15000 to 20000 mA). Fifteen to Twenty Thousand mA is 150 to 200 times the possibly fatal 100 mA mentioned in the above paragraph. OK, this is an extreme example. If it were always true, most of us would already be dead.

The most important concept is:

**DO
NOT
EVER
ALLOW
YOURSELF
TO BECOME
THE GROUND
RETURN PATH!**

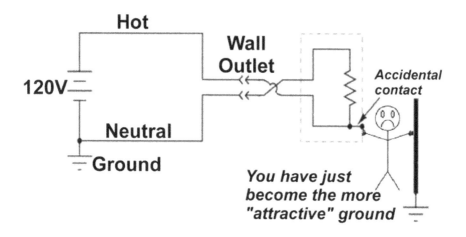

Electricity is kind of like most men. A man will tend to follow (if not actually, at least with his eyes) an attractive woman walking by. Electricity will tend to follow the most attractive ground it finds. One of the basic rules of electricity is that <u>current follows the path of least resistance</u>.

If you and your body are more attractive (the path of least resistance) to the electricity then the previous ground path that it was using, then you will quickly become a <u>new part of the circuit and the new ground return path</u>. Refer to the illustration below.

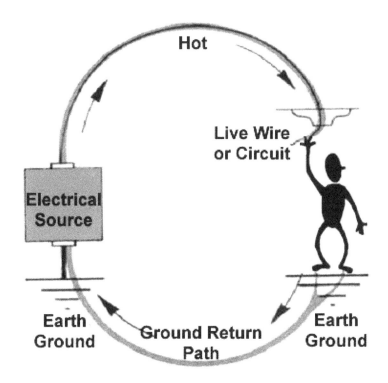

Luckily, in our modern homes with 120VAC wall outlets, smart safety officials have mandated that "Ground Fault Circuit Interrupters" or GFCIs be installed in the house circuits. The basic protection that these provide is that if you do accidentally become a more attractive Ground Return Path, the GFCI will sense that and cut the circuit in microseconds. This usually prevents electrocution.

However, do not disregard common-sense safety practices and depend on the GFCI to keep you from being shocked. Like all things, a GFCI can fail with the result that you might again become a <u>new part of the circuit and the more attractive ground return path</u>.

At all times, troubleshooting should be performed with all power disconnected. This means that you should <u>personally ensure</u> that all the power is disconnected, and that the power remains off while you are working on the circuits You could always have someone else check, but how important is this to him? I mean it is your life at risk, not his...

Lock-outs and signs are commercially available (or can be made locally) that will inform all persons that power MUST REMAIN OFF, and MUST NOT be turned back on until you are finished. This protects both you and your test equipment from damage.

Troubleshooting with the power off is not always possible, and sometimes there is no option but to troubleshoot with the power on. Do this only as a last resort, and be aware that you have just <u>numerically multiplied</u> your risk of being injured.

Think about it and ask yourself, is this the only way to troubleshoot the circuit, and is finding the problem worth losing your life? Or is doing it with power on just quicker and easier? Your Headstone might read something like this:

Here lies
<< Tech John Doe >>
He quickly fixed the problem
And now everything works!
But he doesn't really care
'cuz now he is just dead.

Even after personally locking down the power, it is a good idea to regularly check with a meter that there is no power on the circuit.

Why regularly check? Because there have been numerous instances of other people completely disregarding warning signs, and unlocking and restoring power to circuits under repair. The result can be shocking…. And if you survive, will probably result in the near death of the idiot that re-powered the circuit.

Another common-sense safety practice that most of us *IGNORE* is that we must always remove all articles of clothing or jewelry that could be conductive when working near energized parts. Why anyone would wear jewelry when working on parts, electrical or not, makes no sense to me. Jewelry is way too expensive to damage.

My Experience

When I was first beginning to troubleshoot electrical problems 45 years ago, I was taught a number of "tricks" to use by a wise and old (to me at the time) mentor.

First, approach every electrical circuit as if it is powered. Even if you personally disconnected the power, always treat the circuit as if it has power on it.

I had an incident over 25 years ago where I was working on a 220 VAC 60 Amp line. I followed all the rules. I disconnected the power, locked it off, and posted a sign.

I was confident that the power was off since I had personally turned it off. I began my troubleshooting and all went well for a while. However, at some point, I accidentally touched a contact with my right hand at the same time my left knee (through my pants) brushed against a properly grounded electrical conduit pipe. This should have been no big deal as all power was off.

A close observer said that I seemed to freeze for a couple of seconds, and then just collapsed on the ground. Luckily, no permanent damage occurred.

What went wrong? Apparently, some years before, someone had hardwired 220VAC from another power main into the branch line that I was working on. There was a 200 Amp circuit breaker on the second main power line. The circuit breaker was good, but it did not trip. I guess that means that I got something less than 200 amps of 220VAC thru my body... Lucky for me, because if I had received the full amount I would now be long dead...

So what did I do wrong? I followed the rules, put the sign out, and cut the power. What I did wrong was that I did not verify that there was no power on the line because I was "so sure that there was no power on the line." I did not follow that wise advice "to always treat the circuit as if it has power on it." I was very lucky, not everyone is.

Another trick is always having a safety observer watching you that can help or cut power while you are working around electricity. As you are bouncing around at the end of a live line, you hope the safety observer will save you. Pick someone dependable; do not pick someone that dislikes you.

Still another trick is to keep my left hand in my left rear pocket while troubleshooting. I am right-handed so this works fine. If you are left-handed, just apply this to your right hand. In any case, with my left hand in my left rear pocket, it would be impossible for me to make the "ultimate fatal mistake" of making a Ground Return path from my right hand to my left hand.

I call this situation the "ultimate fatal mistake" because the circuit path in your body goes right through your heart. Since your heart beat depends upon your body's internal electrical stimulation, exposure of your heart muscle to a much stronger outside electrical source is usually fatal.

The more scientific description of this would be that during the Ventricular Relaxation period of the normal heart cycle, the heart is most vulnerable to an external high current pulse. This results in Ventricular Fibrillation and the heart stops pumping blood. This is exactly the opposite of the life-saving Defibrillators that the medics use to start-up hearts.

Back to the story, this trick probably saved my life while I was working on an industrial electrical circuit in a Dry Cleaner Shop in the late 1970's. As most dry cleaner shops are, this one was hot, moist and humid.

I was there to troubleshoot a fault in one of the 220 VAC dryer timer circuits. Unfortunately, there was no "practical" option but to test with power on. There I was, perched on the top of the machine over an electrical circuit sweating like a … I did remember to keep my left hand in my left rear pocket, and all went well… for a while.

Eventually I slipped, lost my balance, and landed on a 220VAC line with my right hand. Because my clothes were wet from all the heat and humidity, I grounded out thru my right pants leg and knee.

While I could not do much to save myself, I did remember not to take my left hand out of my left rear pocket. Instead, I just bounced around on top of the machine until my safety observer cut the power.

I felt like I had just been run over by a truck, but I survived. If I had used my left hand to regain my balance, I would have made the "ultimate fatal mistake" of establishing a Ground Return path from my right hand thru my heart to my left hand. Again, I was very lucky.

Here are the above rules again and I suggest that you memorize them:

1) DO NOT ALLOW YOURSELF TO BECOME THE GROUND RETURN PATH!

2)   ALL TROUBLESHOOTING SHOULD BE PERFORMED WITH ALL POWER OFF!

3) PERSONALLY DISCONNECT, MARK, AND LOCK THE POWER OFF!

4)   APPROACH ALL THE CIRCUITS AS IF THEY ARE POWERED!

5) HAVE AN OBSERVER WATCHING THAT CAN HELP SAVE YOU!

6)   VERIFY BY TEST THAT THE POWER <u>IS REALLY OFF</u>. OCCASIONALLY, TEST AGAIN TO ENSURE THAT THE POWER IS STILL OFF.

7)   CONTINUE TO TREAT ALL CIRCUITS AS IF THEY ARE POWERED!

8) KEEP YOUR OTHER HAND IN YOUR REAR POCKET IN ORDER TO PREVENT ANY CHANCE OF ESTABLISHING A GROUND RETURN PATH THROUGH YOUR HEART MUSCLE.

Any time you are working on or with electricity, appliances, or components, ***ALWAYS EXPECT THE UNEXPECTED!*** Sounds paranoid, but the idea is always to plan a way out in case your personal safety is at risk.

Another Experience

An incident years ago comes to mind when I was repairing and aligning an aircraft weapons turret. I had recently removed a synchro, and was in a hurry to finish up and go home. I had consciously decided to short-cut the alignment procedures and align the azimuth synchro with the system power on. I stuck my arm inside the turret to slightly adjust the synchro. Instead of slightly adjusting it, I accidentally bumped it (mistake #1).

What I did not know is that I had inserted the azimuth synchro 180 degrees backwards (mistake #2). When I bumped the synchro, it immediately commanded the drive motor to move the turret into the azimuth mechanical limits. My arm was now trapped in the turret frame. I tried everything I could think of to get that synchro to let me go, but the only thing I accomplished was getting my arm held tighter in the turret.

After a while, my arm began to hurt. I had only one choice and that was to swallow my pride and start yelling for help. On the positive side, someone heard me yelling my head off, and came and cut the power.

On the negative side, that someone was my Boss (mistake #3).

Some lessons you have to learn the hard way. What did I learn?

ALWAYS
EXPECT THE
UNEXPECTED!

Meter Safety

In addition to your personal safety, there are also safety rules that apply to the use of diagnostic and troubleshooting tools.

Electrical measuring tools or meters (voltmeters, ohmmeters, ammeters, and multimeters) are indispensable aids in your troubleshooting. Safe and efficient use of test meters is an extremely valuable skill for anyone contemplating electrical repair.

This is both for the sake of your own personal safety, and for the continued operation of the measuring tool. Carelessness is the greatest factor when both experienced and beginner technicians to have electrical accidents.

All meters have probes with conductive tips on long wires to allow measurement of electrical signals. It is an ***absolute rule*** that you do not let the conductive probe tips touch one another when they are both in contact with different test points or voltage levels in a circuit.

If this happens on a non-powered circuit, where you are performing resistance or continuity check, then your measurements will be wrong. If you do not notice your mistake, your later troubleshooting may be also be faulty and lead you down the wrong troubleshooting path.

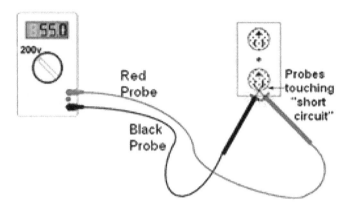

If this happens on a powered circuit with two points of different potential, a short-circuit will immediately occur. Your first sign will be a spark and perhaps even a ball of flame if the voltage source is capable of supplying enough current!

More Meter Safety

Consider each meter to be unique. Just because you "know" one Meter does not mean you can safely operate another. Actually read and understand the Meter operation manual. Pay attention to the Meter specifications and signal limits. Practice with the Meter until you really know how to correctly and safely use it.

Do not attempt to measure more current than the Meter can safely handle (usually around 10 Amps, but each Meter is different.) Do not attempt to measure resistance on a "live" or powered circuit (least – blown fuse, worst – Meter death.) Do not exceed the Meter AC and DC Voltage limits.

Do not hold the Meter probes by the metal tips or use probes that appear to have bad insulation (shocking experience.) And last...

Do not ever become the GROUND RETURN PATH.

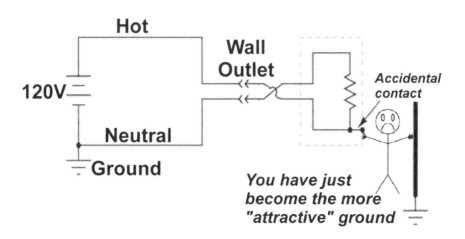

Meter Hints

Below are three meter hints.

1. Make sure your meter is still working correctly as you troubleshoot. Example: An analog needle meter will show zero when it is dead... What if you base your next step on that zero reading? A digital will not show a LCD reading if it is dead, but it sure will show zero if the input lead fuse is blown.

2. If you have to take measurements on two points a long distance apart (more than your lead lengths), do not take the easy way and use the chassis or frame as the return. The resistance of all the rivets and connections will compromise your readings. Use a long external wire as an extension to your lead for more accurate readings.

3. An inexpensive meter may be all you need. Buying an expensive one with all sorts of options that you may never use is not practical. And when you drop it, or burn it up, or leave it in a hot car (and you will) and it dies.... You will not feel as bad losing one that cost you only twenty dollars.

Electro-Static Discharge (ESD)

What exactly is Electro-Static discharge or ESD? ESD is the sudden discharge of the static electricity we build up as we move around.

We experience it everyday. Walking across a carpeted floor in a heated room will build-up enough static charge to give you a noticeable shock if you touch a metal surface. Remember the class joker sliding his feet on the carpet to build up static, and then touching your ear.

While this is not much more than an annoyance for us, that same uncontrolled discharge of static electricity can be fatal to electronic devices or circuits. The ESD does not even have to be great enough for you to notice, and it can still destroy your electronics.

Always check for ESD labels on packages and electronic products. But even if there is not one, assume that it is ESD sensitive. Typical products of daily life such as waxed, painted or plastic surfaces, vinyl tiles, sealed concrete, synthetic materials (Nylon), fiber-glass, finished wood, plastic bags, foam, spray cleaners, solder suckers, and brushes are only a small sampling of what can cause a buildup of static electricity. See the ESD label below.

The Relative Humidity (RH) of an area has a huge effect on the buildup of a static charge. Areas with low RH (10 – 20%) such as the desert will have an extreme ESD danger. Jungle areas with a high RH (65 – 90%) will have moisture and mold problems, but fewer problems with ESD.

Just walking across that carpet in a low RH area can generate 35 thousand volts. That same walk across the carpet in a high RH might only buildup 15 hundred volts. Walking across a vinyl tile floor in a low RH area can generate 12 thousand volts. That same walk across the carpet in a high RH might only build up 200 volts.

A worker moving around at the workbench in a low RH area can generate 6 thousand volts while moving around at the workbench in a high RH might only buildup 100 volts. It is obvious that ESD is a greater problem in low RH areas.

Remember, even 100 volts will destroy some ESD sensitive circuits. ESD control and mitigation has to be one of those "all the time and everywhere" programs.

ESD CCA damage

There are three categories of ESD to a device (most books list two - Catastrophic & Latent with Upset being part of Latent).

1) Catastrophic – that is all she wrote. The device is "kaput" after the ESD event.
2) Latent – Appears to be no damage after the ESD exposure. Lucked out and all is OK. Unfortunately, the life of the component has probably been severely shortened.
3) Upset – This is a temporary loss of the equipment functions. Equipment will resume operation after exposure with no degradation of performance or service life.

Years ago when I was working on a radio, I needed a circuit card assembly (CCA) to replace one that had "smoked." The manual warned about ESD, but when I got the new CCA from supply, it was not wrapped in ESD protective packaging. Oh well; I installed the new CCA. Uh-oh.... No work. It was bad too. Back to Supply for another. Installed it, but it did not work.

Eventually, I realized that these CCAs should have been in ESD protective bags. When I asked the supply people, they told me that the CCAs had come in some plastic envelopes with warnings on the side, but the packages did not fit where they were storing the cards. Their solution was to unpack and throw the bulky packaging out.

This probably would not happen now, since everyone involved in Parts and Maintenance is ESD aware (or should be).

How do you prevent ESD? Complete prevention is impossible. You can minimize ESD by following common-sense rules (these are just a few):

1) Assume all CCAs are ESD sensitive.
2) Handle CCAs as little as possible.
3) Use some type of correctly hooked up anti-static wrist-strap.
4) If you have to handle CCAs, do it by the edges and stay clear of the contacts.
5) Discharge any static electricity from your body BEFORE touching the CCA.

6) Move as around as little as possible; if you do move – Discharge yourself <u>BEFORE</u> touching the CCA.
7) Keep the CCAs in their static protective bags until it is time to install them.
8) Use an air ionizer if available.
9) Avoid bringing static sources near the CCA.
10) Minimize removing or working with any CCAs in a low RH environment.

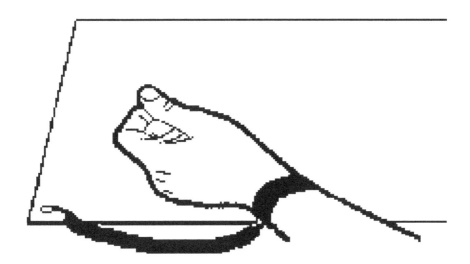

Wrist Anti-static Strap

Nature's ESD

There is one remaining ESD that I need to mention. There is not much you can do to prevent it, but you should be aware of it.

ESD was defined previously as the sudden discharge of the static electricity we build up as we move around. Nature's ESD is lightning. This uncontrollable electrical discharge causes waves of EMI (electromagnetic induction) or EMF (electromagnetic field) to radiate out from the lightning bolt. You have heard this interference noise on the radio or seen the effect on your TV when a lightning bolt strikes close by. Basically annoying, but harmless. Not exactly.

Shielding of wires from EMI protects most modern circuits. However, anything that uses an antenna is at risk. The EMI is a radiated wave or signal, and the purpose of the antenna is to pick-up (receive) or transmit signals.

If the antenna were shielded, than it would not be able to pick up anything. If the lightning is sufficiently close, the antenna will pick up the radiated EMI and create an ESD within the circuit that will damage or destroy the receiver.

Troubleshooting Aids

An understanding of wiring diagrams, schematics, flow charts and trees, visual recognition of electronic components and their respective schematic symbols, and knowledge of some terms (voltage, current, AC, DC, etc) will help you in troubleshooting. Schematics and diagrams, symbols, and troubleshooting trees are covered in this section.

Troubleshooting Charts & Trees

Many electronic component manuals now include a repair section with troubleshooting charts or trees. These trees are developed by the designer, and reflect what he feels will be the most likely causes of failures. They are usually made-up of yes/no or simple questions directing you to the expected fault.

If the design is relatively new, then the fault tree is probably not very reliable. However, if the design has been around for a while with failure reports from the field integrated into it, then the troubleshooting might be quite accurate. Remember that fault trees are almost useless when your fault has been induced by sabotage, or by faulty repairs coming before you.

Troubleshooting Flowchart

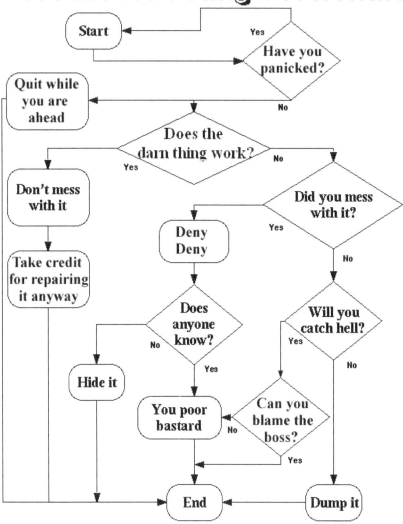

Troubleshooting flow charts and fault trees are made up of four parts. First, the chart has the START. Second, it moves into a series of questions. Third, the questions are answered yes or no. The yes or no answers direct you down the specific path that should solve your problem. Fourth, is the end or conclusion of the flowchart.

The usefulness of the flowchart depends on asking the <u>correct questions</u> that can be answered simply by yes/no. If a question can be answered by something other than yes or no, the question is not specific enough to provide a reliable answer. That is where the problems usually show up.

The designer of the flowchart only asks the questions that he thinks are possible. If a problem comes up that the designer did not foresee, then the flowchart can be misleading.

The goal is to use the tools that you are given, and the flowchart is one of your tools. However, just do not blindly accept the flowchart as gospel... As you are moving thru it, reason out why the questions are being asked, and where the yes/no answers are taking you.

If you find a mistake, or realize that there should be a new question and answer, stop at that point. Starting right there, design your own path using your new questions and yes/no answers. This will adapt that generic flowchart to your specific problem.. and hopefully give you the correct answer you need.

Practice making your own troubleshooting flowcharts. Look at any object, and evaluate its characteristics such as size, color, shape, etc. Design questions and yes/no answers around those characteristics. Move on to an electrical item like a light bulb. Design your own flow chart for it. Maybe start with "Does it light up?" yes or no... Get some practice with it.

Have you ever started troubleshooting something and then been pulled away by a higher priority? You probably intended to get back to your troubleshooting right away, but actually it was months before you got back to it. Where did you stop? If you had written down your steps in a flowchart as you did the troubleshooting, you would know where you stopped even years later....

Schematics & Wiring Diagrams

While we sometimes mix the terms, schematics are used for individual components electronic circuits. Wiring diagrams are simpler (usually) and are used for wire connection information.

Most manufacture operation manuals or technical manuals (TM's) have wiring diagrams of the item component or the system at the back of the manual. Also included sometimes is a connector plug listing.

If a wiring diagram is not readily available, I suggest using the Internet to obtain one. Many are provided free at no charge. Some will have a minimal cost.

If no wiring diagram is available, you can draw one yourself. Be careful... it is extremely slow and a mistake or error can easily destroy the component and your meter, and injure you.

There are rules to wiring diagrams and schematics. Some of the rules are explained in the next section. Remember that rules change as time goes by. And not everyone follows the rules... Big surprise.

Wire Symbols

Some older wiring diagrams and schematics show connecting wires crossing. Non-connecting wires were shown as "jumping" over each other with little half-circle marks.

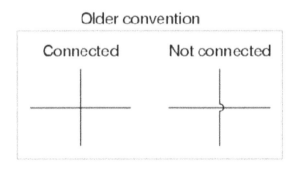

Older convention

Connected   Not connected

Newer wiring diagrams and electrical schematics show connecting wires joining with a solid dot. Non-connecting wires cross with no dot - which is exactly the same as the older convention of crossing wires. Looks confusing, and it is.

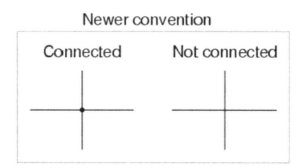

Since there are lots of older diagrams still in use, and lots of older people still use the older convention of connecting wires crossing with no dot, this can and does create confusion and incorrect circuit tracing, and faulty trouble-shooting.

Your best bet is to identify two wires that you are sure connect together. Locate those on the diagram and see how they are shown.

More Symbols

The following are commonly found on electrical diagrams. Note that this list is only a very small sample of what you may find.

Power Sources

DC voltage

DC voltage

AC voltage

Variable
DC voltage

 *A diagonal arrow represents variability for **any** component!*

DC current

Generator

AC current

Resistors

Fixed-value

Rheostat

Potentiometer

Tapped

Thermistor

Capacitors

Non-polarized

Polarized (top positive)

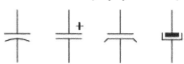

Variable

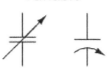

Inductors

Fixed-value

Iron core

Variable

Variac

Tapped

Mutual Inductors

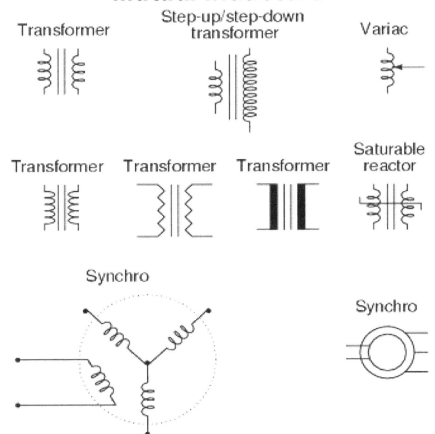

Transformer

Step-up/step-down transformer

Variac

Transformer   Transformer   Transformer

Saturable reactor

Synchro

Synchro

Induction works on the principle that when there is current in one wire, induction can cause current in a nearby second wire that has no physical connection to the first. The most common inductor is the transformer.

Transformers can be step-up where the input (primary) voltage is low and the output (secondary) voltage is high, i.e. Primary = 12V and Secondary = 120V, or step-down where the input (primary) voltage is high and the output (secondary) voltage is low, i.e. Primary = 120V and Secondary = 12V. A good example of a step-down transform is the common wall outlet transformer that powers your computer accessories.

A third type is the isolation transformer where the goal is to keep the two circuits isolated or separate from each other.

The number of windings on the primary input side compared to the number of windings on the secondary output side makes it a step-up or step-down, and determines the secondary output voltage. This number of input windings (or turns) to output windings (or turns) is called the turn ratio.

A simple example is if a transformer has an input or primary of 120V with 10 turns or windings and has a secondary with 1 turn or winding, then it would be a step-down with 10:1 turn ratio outputting 12V. Note that a turn ratio of 30:3 is also 10:1.

The easiest way of figuring your secondary voltage if you know your primary input voltage and the turn ratio is to divide number of primary turns by number of secondary turns, and then divide the input voltage by that number, i.e. 30/3=10 then 120V/10=12V . See the drawing on the next page.

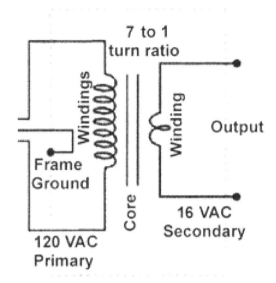

Typical Step-down Transformer

Hand Switches

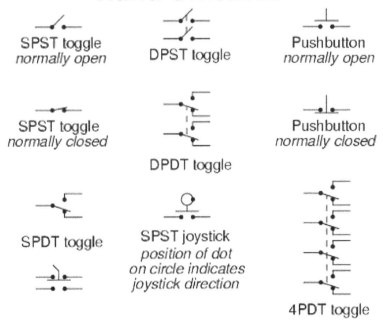

SPST toggle
normally open

DPST toggle

Pushbutton
normally open

SPST toggle
normally closed

DPDT toggle

Pushbutton
normally closed

SPDT toggle

SPST joystick
*position of dot
on circle indicates
joystick direction*

4PDT toggle

The majority of the switches that you will come across are shown below. The dashed vertical lines on the DP toggles show a mechanical connection or ganged between the separate contacts circuits. Turn one on and all come on. Switches are shown in their normal or <u>unswitched</u> position. Normally closed switch position may be shown (NC) while normally open may be shown (NO.)

Automated or Remote Switches

Normally open shown on top; normally closed on bottom

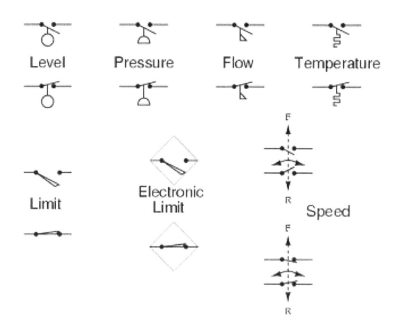

| Level | Pressure | Flow | Temperature |

Limit   Electronic
               Limit   Speed

Relays

Relays require your special attention when you are troubleshooting. One relay can multiply the complexity of a simple circuit by a factor of 6.Whenever I see a relay in a circuit, I approach each terminal of the relay as a separate circuit. I determine whether the relay is Normally Open (NO) or Normally Closed (NC) when the relay coil is not powered.

I also determine at this time whether the wiring diagram or schematic shows the relay in a non-powered state or in a powered state. It can be shown either way, and it is up to you to figure out which it is.

Relay components, "ladder logic" notation style

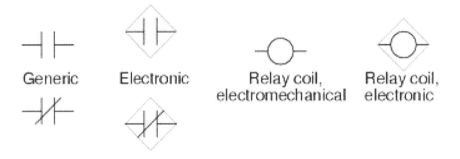

Generic   Electronic   Relay coil,   Relay coil,
                         electromechanical   electronic

Relays, electronic schematic notation style

It may also be drawn wrong. A wrong choice (or assumption) here can cause big troubles in troubleshooting later.

Treat each relay terminal connection as a separate circuit. I follow each circuit back to its start or to its end. I usually start with the input to the relay coil, then the ground to the coil, then each of the commons and the switched contacts.

Do not decide to ignore the coil ground circuit because it is just a Ground. Some newer circuits control everything by switching the Ground on and off.

Connectors (Plugs & Jacks)

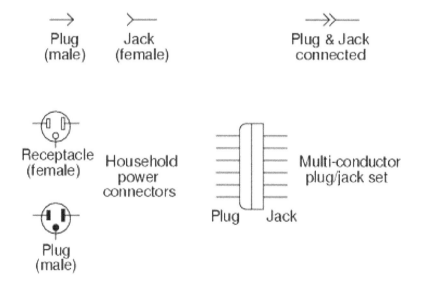

Diodes

Generic	Schottky	Shockley	Constant current

Zener	Light-emitting	Photo-	Step recovery

Tunnel	Varactor	PIN	Vacuum tube

A = Anode
K = Cathode

Transistors

Bipolar NPN	Bipolar PNP	. . . with case

Photo-	Dual-emitter NPN	Dual-emitter PNP

Darlington pair		Sziklai pair
	E = Emitter B = Base C = Collector	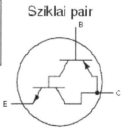

Thyristors

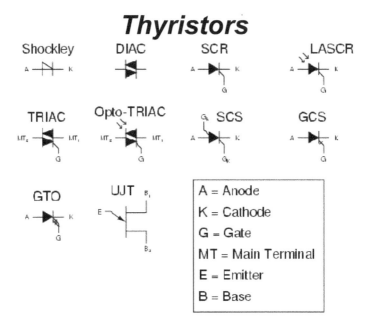

A = Anode	
K = Cathode	
G = Gate	
MT = Main Terminal	
E = Emitter	
B = Base	

Electron Tubes

Vacuum tubes were important devices in the earlier days of electronics. It has not been that long. I know because I started my electronics with tubes. Yes, they had fairly short lives, but they were easy to test and replace. Every grocery store had a tube tester machine that you could use free. It did not take any electronics training to get your black and white TV back on line.

Those days are gone. The only tubes you will see now are CRT or Cathode Ray Tubes in displays, and even those are being replaced by LCDs and flat panels.

Diode
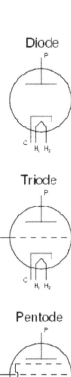

Glow tube
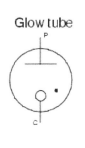

Phototube
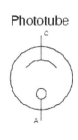

Triode
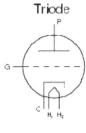

Tetrode
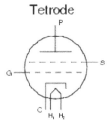

Beam tetrode
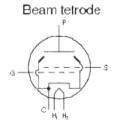

Pentode
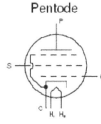

Pentode
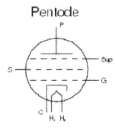

Thyratron
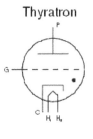

Ignitron
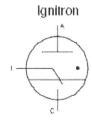

Cathode Ray Tube
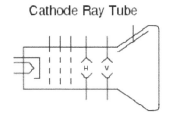

P = Plate	S = Screen
G = Grid	A = Anode
C = Cathode	H = Heater
I = Ignitor	Sup = Suppressor

Integrated Circuits

Operational amplifier   (alternative)   Norton op-amp

Inverter   AND gate   OR gate   XOR gate

Inverter   NAND gate   NOR gate   XNOR gate

Buffer   Negative-AND gate   Negative-OR gate

Gate with open-collector output   Gate with Schmitt trigger input

48

What is Troubleshooting?

Troubleshooting is the act of pinpointing and correcting problems in any kind of system. An auto mechanic uses troubleshooting to determine and repair problems in cars. He uses observations of the car's behavior to decide his approach. OK, nowadays he also uses fault codes and reader.

A doctor does the same with his patients. He observes both the patient and test results to troubleshoot the problem cause. We actually say he is diagnosing the patient and prescribing a cure.

Think of this as Detective or Crime Scene Investigation (CSI) work. Instead of looking for clues to catch the "bad guy", you are looking for clues to catch the "bad cause." OK, it does not sound as cool as being a CSI person, but the theory is the same. Whatever you call it, it is still finding the cause.

Troubleshooters must be able to find the cause or causes of a problem by examining or observing its effects. Often, multiple conflicting and seemingly unrelated observations mask the actual fault source.

Cause and effect relationships are often complex, even for simple systems. An experienced and competent troubleshooter has the ability to identify the root cause of a problem quickly.

Some people seem to be born with a natural talent for troubleshooting; however, it is a skill that can be learned like any other with continued practice.

Sometimes the system to be fault diagnosed is in so bad a condition that there is no logical place to start. In this case, it might be better to just "throw in the towel" and replace it with a new one. Economics and available time will be a big part of that decision.

Usually though, a system is still partially working so that its operation may be tested and adjusted by the troubleshooter as part of a diagnostic procedure.

Here troubleshooting follows the scientific method: determining cause/effect relationships by means of live experimentation.

The scientific method has four steps:
1. Observation and description of a phenomenon or group of phenomena.
2. Formulation of a hypothesis to explain the phenomena.
3. Use of the hypothesis to predict the existence of other phenomena, or to predict quantitatively the results of new observations.
4. Performance of experimental tests of the predictions by properly performed experiments.

My troubleshooting method has similar steps, but I have added a few more to make the process clear:

1. Observation of the system operation and faults.
2. Clearly understand the problem.
3. Gather any additional information.
4. Isolate the problem from everything else.
5. Come up with/formulate/hypothesize a cause or causes to explain the fault observation.
6. Use of the cause or causes to determine the probability of its effect and other expected observations.
7. Performance of experimental tests to determine validity of your suspected cause(s).

The seven steps above will help you come up with answers to the following questions:
1. What actually is the problem?
2. What indicates there is a problem?
3. Is there really a problem?
4. When did this problem occur?
5. What are the possible causes?
6. What is the most probable cause?
7. What is the second, third, fourth, fifth, etc most probable cause?
8. Where do you start your troubleshooting?
9. Which tool(s) should you use to perform the troubleshooting?

Remember the Detective and CSI analogy. You are investigating and looking at everything available for clues as to the cause of the fault(s).

Look everywhere for clues

Once you find the cause, you are only half done. You still have to fix the problem. Sometimes that is as simple as reseating or removing corrosion from a connector, but sometimes it requires replacing that CCA buried in the depths of the faulty component.

While troubleshooting follows rules, it is actually a mixture of those rules and personal creativity. While troubleshooting, you may have to invent your own specific technique adapted to that specific system you are working on. Be creative in examining a problem from different perspectives or angles. Ask different questions when the "standard" questions do not lead to fruitful answers. Some call this "thinking outside the box."

The following troubleshooting concepts are broken down into Questions, Tips, Techniques, and Pitfalls. Some of the concepts are close and maybe repetitive. They are presented here to help you fine-tune your troubleshooting approach.

Troubleshooting Questions

Here are some questions to ask yourself as you are thinking about the cause and effect relationship of a problem:

A) Has the system ever worked before? If not, could it have been installed wrong? Might be time to go back and check the installation or modification instructions for installation errors, or for errors in the directions themselves. Is there another unit just like this one that might also have the same problem?

Years ago, I was tasked to find out why an aircraft Laser distance measuring system was not operating correctly.

I ran all the continuity tests against the Technical Manual schematics, checked connectors, verified correct power, switch operation, and everything else I could think of, and found no faults except that it did not work.

I checked the aircraft next to it and to my surprise found that aircraft system did not work either. So, I checked all 16 of the aircraft and found that none of them worked as described.....

I concluded that there were three possibilities:
1. I did not know what I was doing and my testing was invalid (always a possibility).
2. The same item on all the aircraft had failed and was causing all to be inoperative.
3. The systems were all installed wrong.

Being the overly confident technician that I am, I quickly discarded possibility #1. Looking at possibility #2, it just did not seem likely that all 16 aircraft would have the same failure from a component defect. It was possible, but not likely. That left possibility #3, but these aircraft had been operational for a couple of years.

Was it possible that no one had noticed that this system had not been working on all those aircraft? Or maybe they had noticed, but not felt that it was important enough to write-up. Seems unlikely..... And I had verified that the system was connected exactly as the wiring diagram showed on all 16 aircraft. Time for some serious detective work.

Eventually, I was able to obtain an internal schematic of the main component of the system. I found that the aircraft wiring diagram put power into the control box on an unused connector pin. At least I now had a theory. I temporarily rewired one aircraft, and went out to test it. It worked. I tried the same fix on another aircraft and it worked also.

I submitted my findings and eventually the aircraft manufacturer confirmed that somehow somewhere someone had transposed letters on the wiring diagram and caused the problem. Just one little letter had caused this fault for a whole bunch of aircraft. They were all installed wrong. The moral of this tale is that even the manufacturers can be wrong! Do not assume that what they say (or write) is Gospel.

B) Now, if the system did work before, has anything happened to it since then that could cause the problem? Find out what was the last maintenance performed. Look at the maintenance records, and talk with the maintenance people.

C) Has this system proven itself prone to certain types of failure? Query the manufacturer for failure information. Use the internet for specific failure information on any product. Both of these will give you access to other troubleshooting already performed. You may be able to build on it. Just do not assume it is 100% correct.

D) How urgent is the need for repair? Sometimes, the problem is a minor fault that you can live with. This will allow you ample time to research. Sometimes it is a major fault and needs attention now. This changes your approach.

E) What are the safety concerns before you start troubleshooting? Obviously, you need to be able to do whatever troubleshooting you are contemplating, safely. This means both your personal safety and the safety of the item you are trying to fix. It does no good to find the cause of a minor fault after you have destroyed the item.

F) What are the process quality concerns before you start troubleshooting (what can you do without causing interruptions in production)?

Sometimes the troubleshooter will be required to work on a system that is still in full operation. Once the cause or causes have been narrowed down and verified, then there is the corrective action.

Correcting a system fault without negatively interrupting the operation of a system can be difficult, and sometimes requires thorough planning. When there is high risk involved with the corrective action, it is mandatory for the troubleshooter to plan ahead with options in case of possible trouble.

One other question to ask before going ahead with repairs is, "how and at what point(s) can you stop the repairs if something goes wrong?"

In risky situations, you must have planned "escape routes" in your corrective action, just in case things do not go as you planned they would. That does happen occasionally.

Troubleshooting Tips

As tips, these troubleshooting suggestions serve only as starting points for the troubleshooting process. An essential part of efficient troubleshooting is the assessment probability. These tips may help the troubleshooter determine which possible points of failure are more or less likely than others. Probability is the big factor here. Final isolation of the system failure is usually determined through more specific techniques.

Prior occurrence – Does this device or process have a history of failing in a certain particular way? If the conditions leading to this historical failure have not changed, check for this "way" first. A help to this troubleshooting tip is the rule to keep detailed records of failure. A computer-based failure log is optimal, so that failures may be searched for by time, date, and environmental conditions.

An example of this might be that your car's engine overheated. The last two times this happened, the cause was low water in the radiator. You would immediately check the coolant level first. It might not be the coolant level this time, but it makes sense to check it first.

Past history is no guarantee that the faults are caused by the same problem, but maybe the problem was not really repaired last time. Maybe what you thought was the cause was really just another symptom caused by a deeper fault like a blown head gasket.

On the other hand, if the cause of routine failure in a system has been corrected, then this may not be a probable cause of trouble this time.

Recent alterations - Problems with a system right after some kind of maintenance or other change usually means the problems will be connected to those changes. Usually, but not always.

An example of this could be having a mechanic tune your car engine, and now it misfires. Your first check might be to see whether the spark plug wires were hooked up correctly. Might be that, or it might be something totally unrelated such as computer failure.

Function vs. non-function - If a system is not producing the desired end result, look for what it is doing right. Try to identify where the problem is not. If you can identify all the good circuits, it will leave less of the bad ones to troubleshoot. The components or subsystems necessary for the good circuits to function are probably okay.

An example of this might be a radio that works great on AM, but does not get anything on FM. You can eliminate anything in the radio necessary for the AM band's function from the possible causes.

Obviously, the problem cause is specific to the FM band. If it were part of the AM band, AM would not work either. This eliminates the audio amplifier, speakers, fuse, power supply, and almost all external wiring because they are all part of both the AM and FM. Eliminating parts of the system as possible failure causes reduces the size of the problem.

Look for and record any changes or effects of Hot or Cold Temperatures, Vibration, Moisture – Humidity - Rain, Engines off – Engines on, Aircraft on ground – Aircraft in flight, Idling, Hovering, Reverse motion, Forward motion, In gear, Out of gear, and so on.

Still looking for answers

After collecting and recording everything you can possibly think of:

Read the system or component manufacturers theory of operation. Re-read the theory of operation until you understand it. If you need additional help, try the internet for a simpler explanation. Someone else may have already solved your problem for you. And if they did, and you do not tell anybody, you will look real smart by being able to quickly fix the problem.

This may be the time to honestly evaluate whether you can, or should, continue on this problem, or whether it makes more sense to let someone more knowledgeable handle the troubleshooting. If you just can not understand the theory of operation no matter how many times you read it, then maybe it is time to close the book.

It is possible to continue troubleshooting without a complete understanding of the system operation, but it is a lot more difficult and the probability of frustration is high while the probability of success is very low.

Ok, but you understand the system and now you know how and why the system works as it does. Hypothesize or come up with a theory or theories as to why your system is not working. Based on your knowledge of how the system works, think of what kinds of failures could cause the problem or problems (or observations) to occur. You should be able to come up with a list of possible causes. Do not throw out any that are unlikely. At this point, you should be thinking "outside of the box." Use your imagination.

Now evaluate your list of possible causes and rank them according to the probability that they might happen. Take into account any info on the item's history, or component weaknesses. This is not an exact thing, just put down your best guess and move on. You will be examining the most probable or likely causes first.

An example of this might be that you are happily enjoying life driving down the highway, and all of a sudden the engine starts overheating.

How do you approach this problem? Think about the possible causes for overheating, based on what you know of engine operation (this assumes you know about how the cooling system of your car works; if not, just call a tow truck).

There are really only two possibilities.
1. Either the engine is generating too much heat
2. Not getting rid of the heat well enough (probably more likely).

It is time to Think – Hypothesize –Theorize some possible causes: a loose fan belt, clogged radiator, bad water pump, low coolant level, etc.

Which is the most likely? Which is the easiest to check?

Was there anything that you recently noticed that did not make sense before that might now make sense?

Like that big green puddle on the garage floor as you drove out this morning? That would probably point to low coolant. However, it might not be the "final answer".

Might be that the water pump packing is leaking and it has caused the low coolant level that caused the engine overheating that just warped your aluminum cylinder heads.

You will need to investigate and eliminate each one of your possibilities.

The Flashlight Exercise

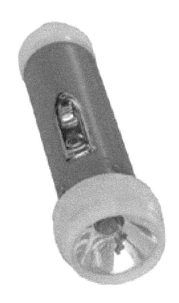

This exercise in thinking "out of the box" uses the typical home two D-cell battery flashlight. First, list all the problems that can keep the flashlight from lighting up (try for at least 30).

This might be easier if you grab a flashlight and look at it. Next, prioritize your list according to probability with the most likely first.

Most people will put "battery" as the very first one on the list. That is a good answer, but we need to be more exact in the failure. Use specific terms like "battery dead," or "battery corroded" or "wrong type battery" for the list.

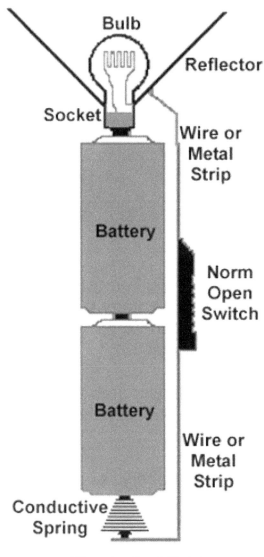

Simple flashlight

Compare your list with my list on the next page.
Here is my list (random probability):
1. battery dead
2. battery missing
3. battery corroded
4. battery wrong type
5. battery installed incorrectly
6. battery contacts bent
7. battery contacts missing
8. battery contacts corroded
9. tension spring missing
10. tension spring corroded
11. tension spring – no tension
12. bulb burned out
13. bulb broken
14. bulb missing
15. bulb wrong type
16. bulb installed incorrectly
17. bulb bottom contact flattened
18. bulb socket loose
19. bulb socket missing
20. bulb socket corroded
21. bulb socket cross-threaded
22. bulb socket contact corroded
23. bulb socket contact bent
24. bulb socket contact missing
25. switch bad - open
26. switch bad - shorted
27. switch corroded
28. switch missing
29. ground wire to switch corroded

30. ground wire to switch open
31. ground wire to switch missing
32. wire switch to bulb socket open
33. wire switch to bulb socket corroded
34. wire switch to bulb socket shorted
35. wire switch to bulb socket missing

Next put your list in what you think are the most to least probable faults. Compare your list with mine below.

Here is my list according to probability:
1. battery dead
2. battery missing
3. bulb burned out
4. bulb broken
5. bulb missing
6. battery installed incorrectly
7. battery wrong type
8. battery corroded
9. battery contacts corroded
10. battery contacts bent
11. tension spring corroded
12. tension spring – no tension
13. battery contacts missing
14. tension spring missing
15. bulb installed incorrectly
16. bulb bottom contact flattened
17. bulb wrong type
18. bulb socket loose
19. bulb socket corroded

20. bulb socket missing
21. bulb socket cross-threaded
22. bulb socket contact bent
23. bulb socket contact corroded
24. bulb socket contact missing
25. switch bad - open
26. switch bad - shorted
27. switch corroded
28. switch missing
29. ground wire to switch corroded
30. ground wire to switch open
31. ground wire to switch missing
32. wire switch to bulb socket open
33. wire switch to bulb socket corroded
34. wire switch to bulb socket shorted
35. wire switch to bulb socket missing

Which would you tackle first? Would it be the most probable, the easiest to check, or maybe a combination of both to decide?

Anything with the batteries would be my first choice. Corrosion would be easy to find with a visual check of all parts, and would eliminate many of the possibilities.

So what good was this exercise? The flashlight has most of the components of larger more complex systems.

If one flashlight has at least 30 possible faults, just think how many electrical fault possibilities there are for a "747" electrical circuit. Even so, remember that no matter how large or complex the system, the approach to troubleshooting is <u>exactly the same</u>.

Find any clues?

Specific Troubleshooting Techniques

After applying some of the general troubleshooting tips to narrow the search for a problem's location, there are some techniques that can isolate it further.

Here are a few:

<u>Are there identical components</u> that are easily to remove and install? Are there parallel or twin subsystems that are the same? If there are, swap the identical components and see whether or not the problem moves with the swapped component.

If it does, you have swapped the faulty component and you are the hero. If it does not, put the swapped one back where it came from and keep looking!

This is an extremely useful troubleshooting method, because it can quickly give you both a positive and a negative indication of the swapped component's fault. When the bad part is moved between identical systems, the original broken subsystem will start working again and the formerly good subsystem will fail.

Sometimes you may swap a component and find that a problem still exists, but has changed in some way.

This indicates that the components you just swapped are somehow different, maybe a different calibration or adjustment. Note the difference for later.

Do not dismiss this information just because it does not lead you straight to the problem. It may be only be coincidence -- look for other changes in the system as a whole as a result of the swap. Think about what these changes tell you about the source of the problem. Remember, you are a detective looking for any clues.

WARNING - WARNING - WARNING

As part of the Mother Nature policy keeping the world in balance, there is a rule that every good thing has to have an opposite bad thing. Like Good and Evil. So here we have an extremely useful and productive tool, so there has to be a catch or an "opposite" thing. There is...

Substituting or exchanging is quicker, but there is a good possibility of causing further damage. Imagine that the bad component has failed because of another unknown failure in the system. You swap the good one in and what happens? Swapping the failed component with a good component causes the good component to fail as well and you are now the proud owner of two pieces of junk.

An example of this might be that you have an electrical circuit that develops a short. This "blows" the protective fuse for that circuit. The blown fuse is not evident by visual inspection, and your multimeter is buried somewhere so that you do not test the fuse for an open. You decide that it is probably just a bad fuse.

To save time, you swap the suspected bad fuse with one of the same rating from a good working circuit. Guess what, the good fuse that you moved to the shorted circuit blows as well.

This leaves you with two blown fuses and two non-working circuits. Of course, it is not all bad because you now know more about the problem. Now you know for certain that the original fuse was blown for a reason.

This knowledge was gained only through the loss of a good fuse and the additional "down time" of the second circuit. Was it worth it? For a fuse, it probably was, but if that was a $5,000 circuit card assembly, you might have a different answer.

As a general rule, this technique of exchanging or swapping identical components should be used only when there is minimal chance of causing additional damage. You have to weigh the probability of damage, the priority of repair, the costs, etc. It is an excellent technique for isolating non-destructive problems, however it will be your butt on the line if you are wrong.

An example of this might be that you have a stereo system that works great on the right speaker, but you cannot hear anything out of the left speaker.

What do you do? Try adjusting the Balance control and see if that helps. There is no reason to start tearing into stuff until after you eliminate the obvious possible causes. Ok, adjusting the Balance Control did nothing.

What do you do next? Try swapping respective components between the left and right channels and see if the problem changes sides, from left to right.

If it does, you have found the defective component. You could swap the speakers between the channels and if the problem moved to the other side (the speaker that was dead is still dead, now that it is connected to the right channel cable) then you know that speaker is bad.

However, if the problem stayed on the same side (the speaker formerly dead now gives out sound after having been connected to the other cable), then you know the speakers are fine. The fault must be somewhere else such as the speaker cable, or speaker fuse, or inside the amplifier.

If the speakers landed up being good, then you could check the cables using the same troubleshooting method. Swap the left cable to the right speaker, and the right cable to the left speaker.

Again, if the problem changes sides (now the right speaker is now "dead" and the left speaker gives out sound), then the cable now connected to the right speaker must be bad.

If neither swap (the speakers nor the cables) causes the problem to change sides from left to right, then the problem might be the speaker fuses in the amplifier. You can swap them also to see if they are at fault. If there is no change no matter what you do, the fault must lie within the amplifier (the left channel output is "dead").

Remove parallel components. If an electrical system has several parallel or redundant components that can be removed without disabling the whole system, you can start removing the parallel components one at a time and see if things start to work better.

An example of this might be if you had a local area network (LAN) that has failed. None of the computers are able to "talk" with each other.

What should you do? You might try unplugging the computers, one at a time from the LAN and see if the network starts working again after an individual computer is disconnected. If it does, then that last unplugged computer could be the one at fault. Possibly it may have been "jamming" the network by putting out continuous noise.

Another example of this could be home circuit breaker keeps tripping open after a short amount of time. This has never happened to me. Yeah, right....

What do you do? Unplug all the appliances on that circuit. Check that circuit breaker does not trip now. If so, plug one appliance in at a time until the breaker "pops" again. If you can identify the problem by plugging/unplugging a single appliance, then that appliance might be defective.

However, if you find that unplugging almost any appliance solves the problem, then the circuit may just be overloaded by too many perfectly good appliances loading down the circuit.

Divide system into sections and test those sections. If the system has multiple sections or stages, you might carefully measure the variables going in and out of each stage until you find a stage where things do not look right or the measurements do not match.

An example of this might be if you had a radio that does not put out any sound at the speaker. What do you do? If you have the knowledge of radio circuit design, try dividing the circuitry into stages like the tuning stage, mixing stages, and amplifier stage. Continue this all the way to the speaker outputs. Measure the signals at the test points between these stages. This should tell you if a stage is working properly. Of course, this requires some test equipment and radio receiver theory knowledge.

Trap a signal. Developments in test equipment have allowed for better troubleshooting over long time periods. It is possible to use instrumentation such as a data logger, chart recorder, or multimeter set on "record" mode to monitor a signal over a long period of time.

This is extremely helpful when dealing with down intermittent problems. Those are the problems that never occur when you test, but show up as soon as you have turned your back and walked away. This may also be useful for monitoring what happens first in a fast-acting system. Many systems have a "first out" monitoring capability that will provide this data.

An example of this might be a turbine engine system that shuts down automatically when it "sees" an abnormal condition. By the time you get to the engine to survey the engines condition, everything is off. It is impossible to tell what was responsible for the initial shutdown, as all operating measurements are now out of the normal range.

What do you do? Video cameras are cheap now. One possibility might be to use one to record what happens by indications on the gauges in an automatic-shutdown event. The videotape playback would show what happened in sequence, down to a frame-by-frame time resolution, and might lead you to the "glitch" that started it all.

Another example of this might be an alarm system that is intermittently going off. Your best guess is that it is being caused by a specific wire connection going bad giving a falsely trigger. Unfortunately, the problem never happens while you are there.

What do you do? Many of the modern digital multimeters are equipped with a "record" mode, which monitors a specific signal such as voltage, current, or resistance over time. It might also tag whether that measurement deviates substantially from a designated standard value. This is an indispensable tool when dealing with intermittent failures.

Failures in Proven Systems

The following things to consider are arranged from most likely to least likely. This order assumes that the circuit or system has been operating as designed and has failed after substantial operation time.

Note that problems in unproven or newly assembled circuits and systems need a different approach. See the next section for that.

Operator error - A too frequent cause of system failure is error on the part of those who are supposed to know how to operate it. This failure cause is at the top of the list and must always be considered first. Verification by the troubleshooter that a fault really does exist prior to tearing the system apart saves time, money, and a lot of frustration.

If operator error is the cause of a failure, it is extremely unlikely that it will be admitted to prior to your troubleshooting. If the operators does not know that he is doing it wrong, he is going to tell you he is doing everything right.

This may only be a training defect, or it may be an unauthorized short cut that has been adopted. Just make sure that you eliminate human error as a cause before tearing things apart. If human error is the cause, be thankful you identified it early on. Try to use your best communication skills in correcting the training defect. No good is accomplished by making the operator feel bad. Remember, you were not always as smart as you are now... Like before you started reading this book.

Bad wire and bad connections. A high percentage of electrical and electronic system problems are caused by one source of trouble, bad wire connections. This includes both open or high resistance, and shorted or no resistance wires. It also includes the connectors in the system. Factors such as high vibration, high humidity, corrosive atmosphere, widely varying temperatures, and battle-damage greatly multiply the effects of this problem.

Connection points found in any plug-and-socket connector, terminal strip, or splice are at the greatest risk for failure. The category of "connections" also includes mechanical switch contacts, which can be thought of as a high-cycle connector.

Improper wire termination lugs such as the old style compression-style connector crimped on the end of a solid wire (very bad) can develop high-resistance as the solid metal "flows" away from the crimp pressure. While the crimp connectors are easy and cheap, solder type connectors are far more reliable.

When troubleshooting connectors, one general rule to remember is that low voltage system connections are usually more troublesome and intermittent than high voltage system connections.

The primary reason for this is the voltage in a higher voltage system will cause an arc across a break in the circuit and blast away insulating (resistance) layers of dirt and corrosion. The arc may even weld the two ends together if sustained long enough.

Low voltage systems generally do not create such vigorous arcing across the gap of a circuit break, and also tend to be more sensitive to additional resistance in the circuit. Note that along with the voltage level, current also plays a role in the amount of arcing in a system.

Open or circuit break failures are more common than shorted failures. However, "shorts" still are a percentage of wiring failures. Many are caused by degradation of wire insulation. This is especially true when the environment includes high vibration, high heat, high humidity, or high voltage.

Many of you may remember the issues the military had with Kapton wiring over a decade ago. While it was lighter and fireproof, some think that it could become brittle and crack because of age, vibration, or movement. I remember removing components and finding wire harnesses with chunks of insulation completely missing with the metal wiring showing.

If you decide to replace old or brittle or suspect wire, make sure that you replace it with the correct type and wire gauge. Wire is identified with an American Wire Gauge or AWG number. The information you need to remember is that as the AWG increases the actual wire size decreases, i.e. AWG 10 wire has 4 times the current carrying area of an AWG 16 wire. This means that if you increase the AWG number of the wire in a circuit you also increase the resistance in that circuit.

You will also increase the total resistance in a circuit if you considerably lengthen the wiring in the circuit. This would require you to move to a larger wire or numerically smaller AWG, i.e. from 14 to 10. A common example of this is the typical outdoor 120VAC AWG 16 extension cord can safely handle 13.5 amps at up to 50 feet in that length. Any AWG 16 cord longer than 50 feet can safely carry only 10 amps.

Shorts may also be caused by conductive buildup across terminal strip sections or the backs of printed circuit boards. This conductive buildup could be caused by exposure to oils, humidity, or even dust.

The most common example of shorted wiring is the ground fault. This is where a conductor accidentally makes contact with either earth or chassis ground. This may change the voltage(s) present between other conductors in the circuit and ground, thereby causing unexpected system malfunctions and/or personnel hazard. This is also another example of what was covered in the Safety section. <u>Do not become the Ground Return Path.</u>

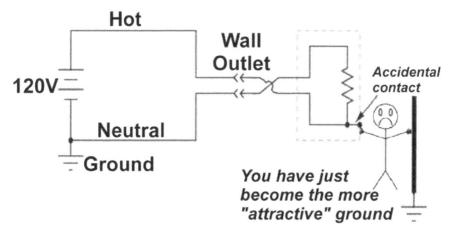

You as the Ground Fault

Occasionally, you may find a relay or mechanical switch shorted. This can only happen in the case of high-current contacts where contact "welding" may occur in over current conditions. More often, the contacts will melt away and become permanently "open".

<u>Power supply problems</u> usually show up as tripped over current protection devices (circuit breakers) or power supply internal damage due to overheating.

Power supply circuitry is usually less complex than the circuitry being powered. Common sense would tell you that it should be less prone to failure. However, because the power supply handles more power than any other portion of the system and operates with greater voltages and/or currents, it is actually more prone to failure.

Active components are amplification devices such as transistors. They tend to fail with greater frequency than passive or non-amplifying devices. This is due to their greater complexity and tendency to amplify over voltage and over current conditions. Semiconductor devices are notoriously prone to failure due to electrical transient (voltage/current surge) overloading and thermal (heat) overloading. The old style electron tube devices were far more resistant to both of these failure modes, but were generally more prone to mechanical failures due to their fragile construction.

Remember that if the manufacturer has provided a socket or connector for the component, than he considers it a part that may fail and need easy replacement. Note that many transistors and integrated circuits plug in to circuit board sockets. On the flip side, if a component is hard-wired making replacement difficult, the designer probably did not expect that it would fail. Of course, the designer could be wrong.

Passive components are the other semiconductors that populate the circuit card assemblies. Their simple designs make them more reliable than the active devices. The following is a list of probable failures:

Capacitors often short out, especially electrolytic capacitors. The electrolyte in a paste form loses moisture with age and eventually fails. The thin dielectric layers may also be damaged by over voltage spikes or transients. Picture below.

Diodes are often used as rectifiers. They usually are found blown open when they fail. However, zener diodes (regulating) are usually found shorted when they fail. Picture on next page.

Inductor and transformer windings will be found open or shorted to their conductive core. Failures caused by overheating or breakdown of the insulation are easily found by the distinctive smell. Picture below.

Resistors are always found to be open when they fail. This happens because as they overheat they actually burn and crumble away. The overheating is usually due to drawing too much current through the resistor. It is less frequently caused by over voltage transient or arc-over.

Note that resistors do not always actually fail when overheated, but may just change their resistance values. This will play havoc with most circuits. A good rule is to visually scan all the resistors looking for a discolored or dark strip around its middle. Look at the examples below.

Connectors are in a world by themselves. They come in all types and sizes. Some are permanent, and some have removable male and female halves.

Connector pin arrangement and numbering can be confusing. Even connectors that look identical may have different pin arrangements. Also, the connector locks may be in different places to make the connectors "Murphy Proof".

Most manuals have illustrations of pin arrangements. If not, consider drawing your own before beginning troubleshooting. It is too easy to probe the wrong pin for a voltage, and then base your troubleshooting on a faulty measurement.

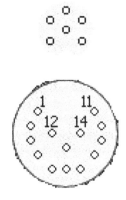

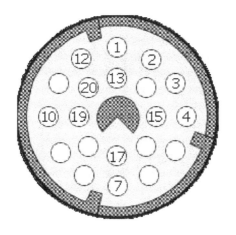

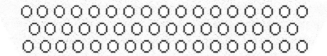

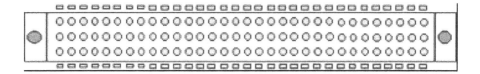

Pin identification can be either numerical or alphabetical. Pay close attention when working with alphabetical letters. There are usually letters that are not use (i,I,l,L), and depending on the number of pins Capital letters, small letters, and double letters (aa, bb, cc) are used.

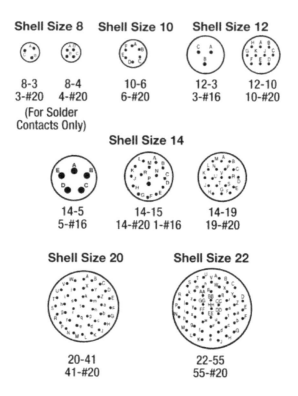

Shell Size 8

8-3
3-#20

8-4
4-#20
(For Solder
Contacts Only)

Shell Size 10

10-6
6-#20

Shell Size 12

12-3
3-#16

12-10
10-#20

Shell Size 14

14-5
5-#16

14-15
14-#20 1-#16

14-19
19-#20

Shell Size 20

20-41
41-#20

Shell Size 22

22-55
55-#20

I have always felt that I have never met a connector I could trust. When you are troubleshooting a problem, each and every connector in the fault line must be physically inspected for correct assembly of both halves, corrosion, bent pins, water or moisture, loose pins, broken insulation, broken wires, and shield strands.

Sometimes the connector problems are self-induced by careless maintenance. Not checking for FOD, sand, or bent pins in the connector before reconnecting is one of the most frequent.

If you see someone putting the female and male halves of a connector together with cannon plug pliers, stop them. The odds are that they have misaligned the halves and already bent some pins.

Shielded wire running between connectors is both a blessing and a curse. The reason to shield wiring is to reduce the risk of outside EMF influence on our low voltage signals. However, working with shielded wire is difficult and even worse, shielding strands that break away from the shielded wire at the back of a connector are almost impossible to see. Just one individual strand is able to ground out a signal on a nearby pin and kill your system.

Moisture conducts electricity and will ground out all your pins. Moisture can form inside connectors as a result of significant weather or pressure changes. Weatherproof and environmental connectors are used, but high pressure cleaning can drive water past the seals. Other liquids also conduct electricity including some spray aerosol cleaners. Make sure that the product you use is non-conductive and leaves no residue.

Loose pins are tough to find. Occasionally, you may find one that will back out of the connector when you put the male and female together. This is because the lock on the pin is broken. Replace the connector.

Bent pins are rarely repairable. Sometimes you are able to straighten them out, but usually they just break off and you have to replace the connector. One option to keep in mind is that if there are unused pins in the connector, you may be able to swap one of them with the bad pin. This method allows you to remove/install only one wire instead of a whole connector of wires.

Environmental splices usually keep the moisture out, but not always. A greenish color is a sure indication of moisture contamination. Sometimes you may find that the splice was not heated correctly and that the solder did not melt. This will cause and intermittent or loose connection.

Failures in Unproven or Modified Systems

The previous section covered component failures in systems that were successfully operating for some time. This section looks at failures in brand-new and modified systems. With these systems, failures are usually related to mistakes in design and assembly caused by people.

Wiring problems. Bad connections are usually due to an assembly error, such as connection to the wrong point or poor connector design or construction. Shorts are also seen, but usually involve conductors accidentally attached to grounding points or wires pinched under box covers.

Another wiring-related problem in modified systems is that of electrostatic or electromagnetic interference induced between different circuits. This is caused by close wiring proximity, and is easily created by routing sets of wires too close to each other. Signal wires are especially sensitive to the EMI of power conductor wires. This condition is very difficult to identify and locate with test equipment.

Systems with built-in-test (BIT) capabilities are sometimes affected by the installation of new wiring. If some of the BIT parameters are based on timing of signals, wires with opposite EMI directions that are too close may delay the signal timing just enough for the timing to be out of the software constraints causing the BIT to trip.

This problem is a killer to troubleshoot. Suggest swapping of components with known good system to isolate problem back to system wiring.

Power supply problems. This is about the same as in the previous section. These usually show up as tripped over-current protection devices (circuit breakers) or power supply internal damage due to overheating. However, the cause usually is one of shorted wiring or an increased excessive load for new components. The actual loads may be larger than expected, resulting in overloading, overheating, and the eventual failure of the power supply.

Defective components. In a newly assembled system, component fault probabilities are not as predictable as in an operating system that fails with age. Any component may be found defective or of the wrong value "out of the box." Never assume that new or serviceable component is in fact good.

A light bulb is a great example. If you had one that was easy to replace, you probably would grab another one off the shelf and just put it in. Occasionally, you find a new one bad and you have to take the new one out and replace it with another new one. However, if the bulb needing replacement was 25 feet up and difficult to reach, you would probably do a continuity check on the new bulb to ensure it was good before climbing up the 25-foot ladder.

I remember an aircraft electrician that was troubleshooting a system. He had identified a bad LRU (line replaceable unit). He got a new one from supply and the system still did not work.

Without doing any testing to verify that the new one was in fact good, he made the assumption that something else was wrong with the system. After weeks of tearing up and testing the rest of the system and finding nothing wrong, a very frustrated electrician called for help.

I was called in and the first thing I did was look at his troubleshooting sequence. I also felt that the original LRU was probably bad, but instead of getting a new one from supply, I borrowed a known good one from another aircraft. There was the risk that I might destroy the borrowed LRU, but that did not happen.

In fact, the system worked perfectly. All that time and maintenance wasted, all that frustration created, because of continued troubleshooting based on an incorrect assumption.

Improper system configuration. Components with the wrong values may be used in a new system. Resistors with watt or power ratings too low or tolerances too large may have been installed in the system. Sensors, instruments, and controlling mechanisms may not have been calibrated, or were calibrated to the wrong ranges.

Design error. No designer ever wants to hear from the field that his design is wrong. If he does hear it, no designer wants to believe that his design is wrong. Usually, it is a no win situation for the field person.

He has the problem and the suspected reason for the failure, but the back-at-home designer has all the scientific proof and reputation that says the field person is "nuts" or incompetent.

While most design goofs show-up early in the operational life of the system, some do not appear until just the right (or wrong) conditions exist. Design flaws are the most difficult to uncover, as the troubleshooter normally ignores the possibility of design error because the system is assumed to be "proven."

I recently did a "chip" (integrated circuit) change and software up-grade on a very high-end computerized sewing and embroidery machine. The machine worked perfectly before the up-grade, and seemed to work perfectly when I was finished.

Over a few days, a problem surfaced. It would not turn on right away on cooler days. I got it back and went thru everything again, i.e. checked all connections, wires, etc. Everything tested perfectly and I gave it back to its owner.

A few days go by with same problem. Frustrated, I got together with the owner to try different ideas. We discovered that if the embroidery unit was connected the machine would start right up the first time and every time. If the embroidery unit was not connected, it might take 10 minutes and 10 times to start it up.

There are only three possible causes for this:
1. I really messed up the installation… but if that were the case, other things on the machine would not work correctly. Everything else did… even the new functions (not very likely since it works perfectly once it starts).
2. The chip is defective (not very likely since it works perfectly once it starts).
3. There is a software problem….. This is the most likely, and is a design error. Other than reporting it, there is nothing I can do until a new upgrade comes out.

A problem like this can be extremely frustrating. It seems like it is a hardware problem, but in reality is software generated.

Causality made me think it might be related to my physical changing of the IC. This caused me to disassemble everything again to recheck my work. It was actually "Coincidence" as far as the chip replacement and my service. However, it was "Causality" as far as the software up-grade.

Confused.... the lines can get real blurry. Remember to "think out of the box".

Troubleshooting Pitfalls

Some of these have been covered before, but are important enough to be covered again. Faulty reasoning is the main reason for more failed troubleshooting efforts than any other impediments. Following are a few more common troubleshooting mistakes.

1) Trusting that a brand-new component will always be good. While generally true, this is not *always* true. It is also very possible that a component has been incorrectly labeled. It may actually be a non-serviceable one, or it may be a slightly different edition of the same model. If possible, installing a known good component to test the system is the recommended procedure. Then if the system works, you can replace the borrowed one with a new one from supply. If the system then does not work after the new one from supply is installed, it is a pretty open and shut case as to the condition (bad) of the new one from supply. (One exception - LRU tolerances... see page 101 #9).

2) Not periodically checking your test equipment. This is especially true with battery-powered meters, as weak batteries may give incorrect readings. Remember to test the meter on a known source of voltage, both before and after checking the circuit to be serviced, to make sure the meter is in proper

operating condition. This takes a little more time, but it can save hours chasing a voltage or resistance problem that does not exist.

If you are using an analog (needle type) multimeter, the very first thing you should check after turning it on is the needle "Zeroed." What does that mean? Look at the dial and needle and make sure the needle is over the zero line. If not, adjust it. Also, each time you change a scale (x10, x100, 1000, 1V, 10V, etc.) or signal type (AC, DC, Ohms, Amps, etc) you need to recheck to make sure the needle is zeroed.

3) Assuming there is only one failure contributing to the problem. Single failure problems are great for easy troubleshooting, but sometimes causes come in multiple numbers. Sometimes the failure of one component may lead to a system condition that damages other components. Sometimes a component in marginal condition goes undetected for a long time, and then when another component fails the system suffers from problems with both components.

A good example of this is the modern auto computer system. It used to be that cars needed tune-ups every 6000 miles in order to keep them running. Now we have computers that compensate for the wear on individual components and this gives us extended maintenance schedules. Everything works longer until the computer can no longer compensate, and then you have multiple worn-out components creating a system failure.

4) Mistaking coincidence for causality. Just because two events occurred at or nearly at the same time does not necessarily mean one event caused the other. The events may be both related to a common cause, or they may be totally unrelated and coincidental.

Try to duplicate the same condition suspected to be the cause and observe the results. Do you get the same results? Did the event suspected to be the coincidence happen again? If not, then there are two possibilities.

A) There is no causal relationship and it was a coincidence.
B) You did not duplicate the condition correctly.

5) Self-induced blindness or "Too Close to the Forest to See the Trees." After a long day of troubleshooting a difficult problem, you will become tired and most likely begin to overlook crucial clues to the problem. Take a break and let someone else look at it for a while. You might be happy at the difference this can make.

Usually "team troubleshooting" takes more time and causes more frustration than doing it yourself. An exception to this is when the knowledge and the investigative approach of the troubleshooters are complementary.

6) Failing to question the troubleshooting work of others on the same job. This just sounds wrong, because no one wants to have his or her work questioned. Because it is easy to overlook important clues or details, troubleshooting information received from another troubleshooter should be personally verified before proceeding. This is a common situation when troubleshooters "change shifts."

It is important to exchange information, but do not assume the prior technician did check everything reported, or that they did it right. I have had to backtrack many times because I assumed and failed to verify what someone else told me.

7) Being pressured to "hurry up." Any time an important system fails; there will be pressure from other people to fix the problem as quickly as possible. Pressure does not help anyone, and sometimes results in further damage. Most failures can be fixed or temporarily repaired in short time if approached logically.

Bad "fixes" resulting in haste often lead to damage that cannot be ever fixed. If the potential for greater harm occurs, the troubleshooter needs to maintain his logical approach in the midst of chaos. Going off "like a chicken with its head cut-off" benefits no one.

8) Finger-pointing. It is too easy to blame a problem on someone else. The goal is to fix the problem and learn how to prevent the problem from ever happening again (if possible.)

9) LRU (line replaceable units) Tolerances. All components have tolerances. While a component may pass an individual self-test, it may not pass a system self-test on a particular aircraft. The component may still work in another aircraft, but just does not work with the other components in that one aircraft. It would still be considered a serviceable part.

Ok, maybe right now you are asking yourself what am I talking about. The simplest way to explain this is to imagine that you have a complex electronic system composed of many replaceable units. Let's say that these units come out of the factory in three tolerance levels of +, -, and zero.

Normally, a system would be a random makeup of components with all three tolerance levels of +, -, and zero. What would happen if a particular system landed up with all + components, or all – components? Most likely, the overall system would exceed the total system allowances and the system would not operate correctly.

An example of this repeatedly occurred when I was working with an aircraft missile system. Thru trial and error we learned that while some components would not work in a particular aircraft, they were perfect in another. This also meant that if you had an aircraft that would not work with a new unit out of supply, you would be able to get all the aircraft operational if you swapped (controlled exchange) among four or five aircraft. Still confused, email me.

11) Compatible Designs. We tend to want to upgrade and modernize everything to get the most for our money. Sometimes, things are not backward compatible. If you are lucky, they can be modified to work.

Many years back, I was updating an aircraft from a Black-White (BW) Weather Radar to a Color Weather Radar. The two radars were 20 years apart in technology.

I followed the installation instructions exactly, and it actually did work; it just did not work right. Some type of interference was distorting what was on the screen. Back to the "drawing board" where I checked and re-checked every wire, connection, and shield and found absolutely nothing wrong or abnormal. I called the manufacturer and explained my problem, and they were as baffled as I was. I exchanged units and antennas with no improvement.

One day, frustrated, I just sat there and stared at the old BW unit and the new Color unit sitting on the table. The Boss had me remove both from the aircraft, and was about to throw them and me out of the building. As I looked at the units, I noticed how much they had changed over the years, especially the antennas. The old BW one was large and curved, and the Color one was small and flat. All of a sudden, a light went on in the dark recesses of my brain. Maybe because of the antenna shapes, the systems were not "seeing" the same thing.

To protect the units from the weather, the antennas were covered by a fiberglass radome that was designed to be invisible to the Weather Radar signal. My new theory was that the new Color Weather Radar antenna shape changed something and caused the fiberglass radome to become visible to the Color Radar.

I decided to share my idea with my coworkers. It brought many laughs from them. Ignoring them, I came up with an idea to space the Color Weather Radar antenna out from its mounting plate so that it had about the same arc of travel as the old BW Weather Radar antenna. I installed the spacers,,,, Voila! It worked perfectly.

Ok, so what was the cause of the problem? Because the arc of travel of the Color Weather Radar antenna was different, its radar signal no longer passed through the radome fiberglass material at a 90-degree angle as the BW Weather Radar signal did. This caused the Color Weather Radar Signal to return off the fiberglass material in the radome before the signal ever left the aircraft. Visually, this showed as signal interference.

11) Induced Failures. This refers to troubleshooting by someone going before you. Instead of fixing the original problem, they added more or induced faults that make it difficult to see the original fault. Poor recording of the troubleshooting already performed makes this worse. Always keep a written record of what you do in your troubleshooting.

While "in the sand" in Baghdad, I was asked to look at an aircraft with a stabilator failure. The system had been tested repeatedly with test set. Unfortunately, there had been no success in identifying the cause.

This was one of those times when I was coming in after a whole bunch of other technicians had already attempted to fix the system. Luckily, I had another troubleshooter with me that knew as much as I did about the system. Craig and I made an excellent team.

We immediately looked at all the previous maintenance. Since all the replaceable units had been replaced many, many times with no effect, we decided to look at the wiring.

After running all possible wiring for both high/low resistance and resistance to ground, we found two wires that were shorted to ground that should not have been.

We examined the wires, and found that some time in the past someone had decided to repair and rewire the connections. Accidentally, they had connected the output of the first unit that was a monitored input to the second unit to the shield ground, and connected the output of the second unit that was a monitored input to the first unit to the shield ground. This simply grounded the outputs of both units.

When the wiring was repaired and connected properly, the system returned to operational. As usual, no one admitted to ever rewiring that connection. It was magic….

12) Purposely Induced failure or SABOTAGE. I have saved this for the last because there is no troubleshooting more difficult and frustrating than troubleshooting a problem that is the result of sabotage.

If you are troubleshooting a problem in a system that has been caused on purpose by another troubleshooter that is knowledgeable on the system, you are in for a wide-awake nightmare.

I have already mentioned the fact that system test sets are not usually adequate to diagnose induced failures. Test sets normally test for problems that wear and tear creates, or for known and expected system failures. Possible faults such as two wires pulling out a plug and then being soldered together logically cannot happen (but they do). If a test set was designed to test for all possible defects (including induced), then it would take 10 times as long to test and be huge in size.

Troubleshooting of sabotage to a system by a knowledgeable technician basically will come down to which of you really "know" more about the system. In this case, remember to 1) assume absolutely nothing, 2) nothing is as it may seem to be, 3) keep meticulous records of everything you do, and 4) verify everything again and again.

I have two examples of sabotage that I have experienced. Both were major learning lessons for me.

A) The first example involves something that many troubleshooters have experience with. It was a fairly simple aircraft radio system with a transmitter and receiver, control head, antenna, and associated wiring. The fault recorded was no transmit and poor reception.

By the time I was called in, all the replaceable units had been replaced numerous times with zero success in finding the problem. I triple checked all the previous maintenance and could find no fault other than the previous technicians kept replacing the same components in a vicious circle.

The system appeared to be operating normally except that it did not transmit and would receive poorly. I zeroed in on the antenna line. The previous troubleshooters had run a continuity check of the coax antenna line from center wire to shield, and had gotten continuity. Since they had not removed the coax from the antenna, they figured they were reading the internal antenna loop resistance.

They did remove the coax line from the antenna and check for high resistance from one end to the other for the center wire and for the shielding. The antenna coax line checked out OK, so they reassembled everything. They were back to "square one" with no answers.

Suspecting that it was an antenna coax problem, I had a few alternatives at this point. 1) I could run an entirely new temporary antennal coax line outside the aircraft and check the operation, 2) recheck their troubleshooting of the antenna coax line, or 3) do a transmit SWR check (Standing Waves Ratio) and look for reflected power.

To me the simplest alternative was #2. I removed the antenna and isolated the antenna coax line before performing any checks. I verified that the resistance readings from end-to-end for both the center wire and the shield were correct. However, when I checked for continuity from the coax shield to the coax center wire, I found almost a "dead short". The previous troubleshooters had assumed that the internal antenna loop resistance caused this reading. That was their mistake.

The next step was to find where the "dead short" was. Common sense told me to check the connectors for shield strands and insulation breakdowns. No luck; all connectors were A-OK. Feeling frustrated, I even removed the connectors and personally installed new ones with no change in the fault.

My next step was to visually inspect the full run of the antenna coax line for possible chaffing or damage. You never know, someone may have inadvertently drilled into the line. A visual inspection of the complete line gave me no knew info.

As the frustration level increased and my confidence level decreased, I decided to change tactics. I went back to alternative #1 and took the time to run an entirely new temporary antennal coax line outside the aircraft and check the operation. Guess what? With the new temporary antenna line, the system transmitted and received OK. This proved I was on the right track all the time.

While my confidence was restored, I still had to fix the existing antenna coax line. I didn't think that the pilots would be too keen on flying around with the new antenna line taped to the outside of the aircraft with 100 mile-per-hour tape. I didn't see any problem with that.

All this happened over 15 years ago. If I ran into this problem today, I would use a TDR (Time Domain Reflectometer) to tell me how far down the antenna cable the defect (short) was and the problem would be solved quickly.

Back then, I had to do the "touchy-feely" test. That is where you run your fingers across all the coax from one end to the other feeling for a bad spot in the cable. I finally found something wrong in the tail boom section. Upon close examination, I found that a straight pin had been pushed through the outside of the coax thru the shielding thru the insulation hitting the center wire and the out the other side. It had been cut off on the other side to make it harder to find.

I removed the straight pin and repaired the coax cable section. System testing verified correct transmission and reception, and the problem was fixed. We never did figure out who was responsible for sabotaging the coax. Troubleshooting a success – Fixed– Fully Mission Capable.

B) The second example was even more years back. I asked to look at a problem with an aircraft missile system. Just like the first problem above, I was coming in after many other troubleshooters, and all the replaceable items had been replaced.

Normally, verification that the aircraft missile system was operational would be performed with a dedicated semi-automated missile system Test Set. The problem noted in the maintenance log was "system makes test set internal circuit breaker pop immediately upon power on". This pretty well eliminated any use of the Test Set for troubleshooting help.

I double-checked all the previous maintenance and could find no fault other than they kept replacing the same components. Since the Test Set would not stay on-line and was useless, I was going to have to do this the old fashioned away with the system BIT (built-in-test) and a multimeter.

After disconnecting the Test Set, I ran the system manually. BIT failed right away, but no aircraft or system circuit breakers popped. The problem with the Test Set sounded like a "short", but if that was true then I should have had an aircraft or system circuit breaker pop when I operated the system manually. That did not happen.

Maybe it was two problems, 1) an aircraft problem, and 2) a bad Test Set. I self-tested the Test Set and found that it was working properly. Ok, what now? Maybe it was something the Test Set self-test didn't test. I borrowed another "known good" Test Set from another unit. Same thing, the Test Set circuit breaker popped as soon as I turned the missile system on.

Frustration level increasing, I realized that I had made a mistake. I was using the original Test Set cables to hook-up the new test set to the aircraft. It was probably a cable problem. I replaced all the old test cables with the new Test Set cables and using the new Test Set attempted to power up the system. No Go, circuit breaker still popped.

Supervisor wanders by and innocently asks me what is taking me so long? My response is not printable. Frustration level increases by ten-fold. I take both Test Sets and both cable sets and hook them up one at a time to an aircraft with a "known good" missile system and test. No circuit breakers pop, both the aircraft missile system, and both Test Sets and wire sets operate perfectly. I had now spent a lot of time and accomplished almost nothing.

I had learned that it was an aircraft problem and not a Test Set problem. But if it was an aircraft problem, how come the aircraft missile system circuit breakers didn't pop? Maybe it was in a circuit in the aircraft that was not an operation circuit, but was part of the test circuit. I spent the next 2 weeks running continuity and ground checks on what seemed like 10 thousand wires while having my Supervisor wander by and repeatedly ask me what is taking me so long? I found nothing.

Frustration level now at maximum. Here, I had a multimillion-dollar aircraft that had been sitting in the hangar for three weeks and I was no closer to fixing the problem than when I started.

What to do, what to do? Visual inspections, BIT, continuity and shorts testing told me nothing. It was definitely time for an "out-of-the box" experience. Since the only problem clue was the Test Set internal circuit breaker, I decided to take another approach. I researched the inputs and found a number of possibilities. After thoroughly checking all the possibilities, I knew that there was a problem close, but where? What do, what to do?

I decided to try one last thing before I "went down in flames" and had to give up. (Note: <u>I do not in any way ever recommend the following procedure.</u> It is only included here to illustrate troubleshooting of sabotage.) I needed to see if I could trigger anything that would help me zero in on where the problem was, so I disabled the Test Set circuit breaker (so that it could not pop). The worst that could happen was that I would destroy a quarter million dollar Test Set.

I turned on the Test Set and all went well for about 3 seconds when smoke started boiling out of the aircraft electronics opening. Where there is smoke, there is fire (or at least some thing charred). I finally had the visual clue I was looking for, and it was a small success. An elated feeling came over me.

Unfortunately, the feeling was very short-lived. The smoke had caused a fire alarm, the hangar was evacuated, and a fire truck arrived. The supervision chain did not share "my elated feeling" or the opinion that I had used the correct troubleshooting procedure. In fact, there was talk of holding me monetarily responsible for all damages and costs. They left me with the warning "that they would get back to me."

Ignoring little things like money, I proceeded with my troubleshooting. I had now had a helper (or guard) whose mission was to keep me from doing any more damaging or careless (stupid) things. "ye of little faith".... I quickly found where all the smoke had come from. The distinctive smell and charred remains of two current limiting resistors showed exactly which circuit had a problem.

With those visual cues and circuit identification, I was able to locate the wire runs that were suspect. Using the "feely-touchy" method, I eventually found a break in a wire near the front of the aircraft. The strange thing was that to get to that wire <u>I had to cut 10 wire ties</u>, and then the wire was <u>in the very center of a group of over 50 wires</u>. How did it get damaged?

Using a magnifying glass, I was able to see that the wire had been cut with a diagonal wire cutter. I repaired the wire with an environmental splice, and replaced the circuit card assembly that I had charred.

Time to see if I fixed it or not... BIT and Test Set testing showed the missile system was now fully operational. The strange thing was that there were no other repairs in that area of the wires, and all the other wires had been wire-tied around the cut one. Looked like sabotage to me.

Of course, the outcome was that aircraft and missile system was now operational. I received hearty congratulations for the results, and a stern warning from my non-technical supervisors not to do it that way again.

They had no idea how I should have done it, they just knew that I did it all wrong. Honestly, even today (over 30 years later), I am not sure how I could have done it any other way. Sometimes you win, sometimes you lose, and sometimes you're just lucky to stay even...

Multiplex/Mux Bus Systems

This section will explain only the very basics of Multiplex or Mux Bus systems. Anything more than that is beyond the scope of this book. More detailed information is available free on the Internet. For specific Mux Bus information used in the design of a commercial component, contact the manufacturer.

The Mux Bus is a shared communication link between components or subsystems. While it is more difficult to troubleshoot with standard test equipment, it is widely used for the following reasons. Low cost: a single set of wires connecting components, Versatility; easy to add more components, Standard: components designed the standard may be used on completely different equipment, i.e. "plug and Play", and Weight Reduction: One or two sets of wires may replace hundreds of wires reducing overall weight.

How does it work? When I first started teaching Multiplex theory on military aircraft twenty-five years ago, it was much harder to explain. I used the Post Office explanation... When you mail a letter thru the Postal System or send a package thru FED EX, you send them to a specific address, i.e. number street-city-zip code. You also put on your return address with number street-city-zip code. Both the origin point (you) and the destination have unique addresses. Think of the Postal System as a giant Multiplex system where each house is being a component of the system. Each house has its own unique address that ensures it receives and sends correct communication (mail).

The electrical multiplex system is the exactly the same. Each component (house) has its own unique electrical (data) address that ensures it receives correct communication. Because each component (house) address is unique, the data communications only arrive at the correct component.

A more modern explanation involves your computer. If you have a DSL or cable or satellite you are using a Multiplex system. When you use them, your computer is identified at all times by its unique electronic address. This keeps your search requests, emails, uploads, or downloads separate from all the millions of other search requests, emails, uploads, or downloads or data bits that are moving along the Internet at the same time. Your computer address also identifies you thru "cookies" to commercial entities…. And to hackers.

How do you troubleshoot it? I suggest component substitution, while continuity checks still work for the Mux bus system wiring. Those are the basics.

Troubleshooting Exercises

Using our trusty 2 D cell flashlight again, let's look at how we could verify or eliminate each one of the faults we previously identified.

1. battery dead –measure output with voltmeter
2. battery missing – visual insp (inspection)
3. bulb burned out – measure continuity
4. bulb broken – visual insp (inspection)
5. bulb missing – visual insp (inspection)
6. battery installed incorrectly – visual insp
7. battery wrong type - visual inspection, measure output with voltmeter
8. battery corroded - visual inspection
9. battery contacts corroded - visual inspection
10. battery contacts bent - visual inspection
11. tension spring corroded - visual inspection
12. tension spring – no tension - visual inspection
13. battery contacts missing - visual inspection
14. tension spring missing - visual inspection
15. bulb installed incorrectly - visual inspection
16. bulb bottom contact flattened - visual inspection
17. bulb wrong type - visual inspection
18. bulb socket loose - visual inspection
19. bulb socket corroded - visual inspection
20. bulb socket missing - visual inspection

21. bulb socket cross-threaded - visual inspection
22. bulb socket contact bent - visual inspection
23. bulb socket contact corroded - visual inspection
24. bulb socket contact missing - visual inspection
25. switch bad – open - measure continuity
26. switch bad – shorted - measure continuity
27. switch corroded - visual inspection
28. switch missing - visual inspection
29. ground wire to switch corroded - visual insp
30. ground wire to switch open - visual insp; measure continuity with ohmmeter
31. ground wire to switch missing - visual insp.
32. switch to bulb socket open - visual insp; measure continuity with ohmmeter
33. switch to bulb socket corroded - visual inspection
34. wire switch to bulb socket shorted - measure continuity with ohmmeter
35. wire switch to bulb socket missing - visual inspection

It appears that the majority of the faults could be eliminated by visual checks or inspections, with a few needing either voltage or continuity checks.

The next section has four simple circuits plus two complex circuits for you to troubleshoot

Diagram Example 1

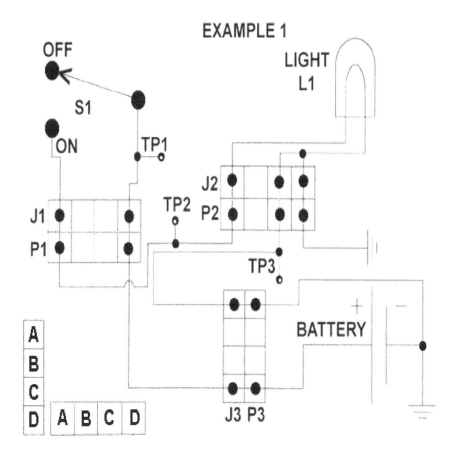

EXAMPLE 1

This is a representation of a circuit consisting of a 6-cell battery, connectors, SPST (single pole single throw) switch, light, and three test points. Use the connector pin layouts on each wiring diagram to identify the correct pins.

The circuit operation follows: Positive voltage from the battery follows the wire thru connector J3/P3 pins D, then down the wire thru connector J1/P1 pins D past TP1 to the wiper common point of No (normally open) switch S1.

Normally open means that the switch is a "break" or open in the circuit. When the switch is "turned on" and closed, the circuit will be completed.

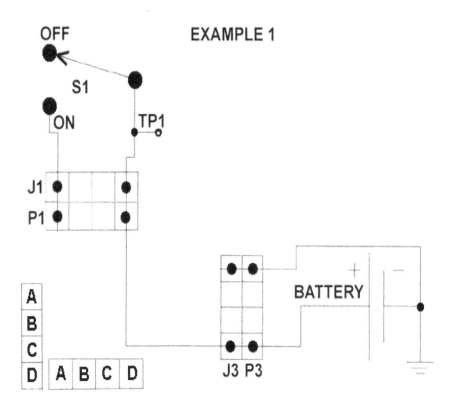

Note TP1 also connects here. This is one of a number of test points that allows the troubleshooter easy access to the circuit. This allows you to take your measurements without disturbing the wiring or any protective coatings on circuit cards.

If switch S1 is closed, positive voltage follows the second wire thru J1/P1 pins A (note that TP2 also connects to this line), then down the wire thru connector J2/P2 pins A to the L1 Light.

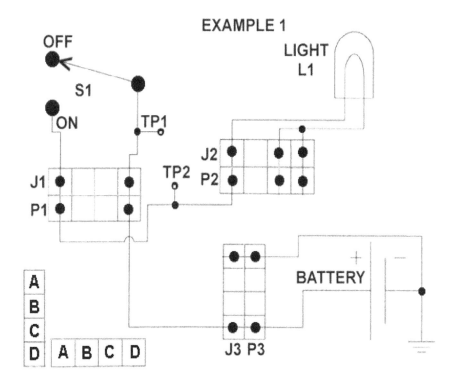

EXAMPLE 1

With Positive voltage present, the L1 light filament will glow if the circuit has a return ground path (and if the bulb is good).

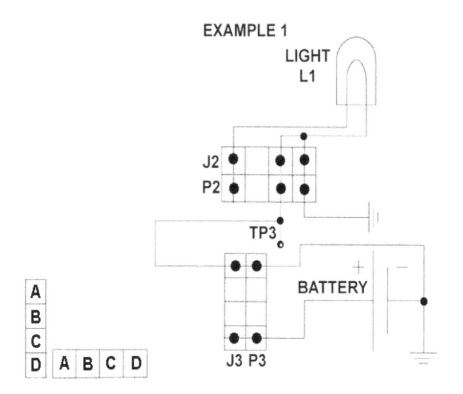

EXAMPLE 1

The return path is from the Negative side of the filament thru the other lead thru a common tie point thru J2/P2 pins D and then to a chassis ground.

An alternate path from the tie point is thru J2/P2 pins C past TP3 thru J3/P3 pins A to the battery negative and ground.

What would you use, and how would you troubleshoot this circuit? Assume that the bulb does not light no matter what position the switch S1 is in.

There are two methods you can use, either with power on by measuring voltage, or by measuring continuity and resistance with power off and disconnected .

EXAMPLE 1 TROUBLESHOOTING

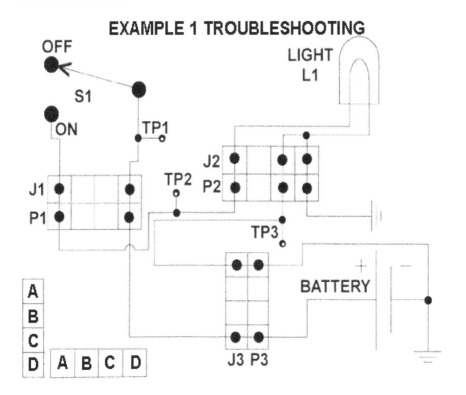

Power On - Using a multimeter, (after verifying correct operation of meter), establish that the battery has a charge by measuring voltage from the positive terminal to the negative terminal. Assume for this explanation that you get 6VDC.

Keeping your negative (black) probe on the – battery terminal, touch your positive (red) probe to TP1. The meter should show 6VDC.

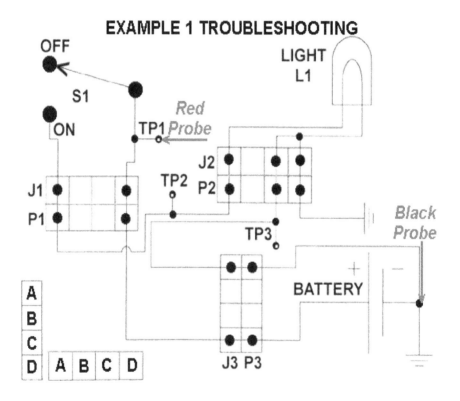

EXAMPLE 1 TROUBLESHOOTING

If it does not, then there is a circuit disconnect in the wiring or connectors between TP1 and the positive battery terminal. Disconnecting J1/P1 and probing pin D for the 6VDC, and if not present, disconnecting J3/P3 and looking for the 6VDC at pin D should show you where the break is.

Assuming you have 6VDC at TP1, move your + red probe to TP2 and check for voltage. You should have zero VDC.

Close (turn on) switch S1 and you will have 6VDC at TP2. If not, the problem is either with switch S1, J1/P1 pin A or the wiring to TP2.

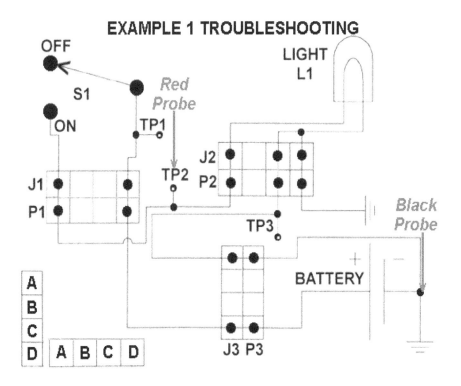

EXAMPLE 1 TROUBLESHOOTING

Assuming that there is 6VDC at TP2, check for 6VDC at J2/P2 pin A. That completes the power side.

Let's look at the − or ground side of the circuit. Everything on the output side of the light is the ground side. To measure that using power on, we need to switch the position of our test probes.

Touch the + red probe to the positive terminal of the battery. Touch the – black probe to the two chassis grounds and TP3. All should show you 6VDC. If not look for breaks in the wiring, and disconnect J3/P3 and probe pin A if required.

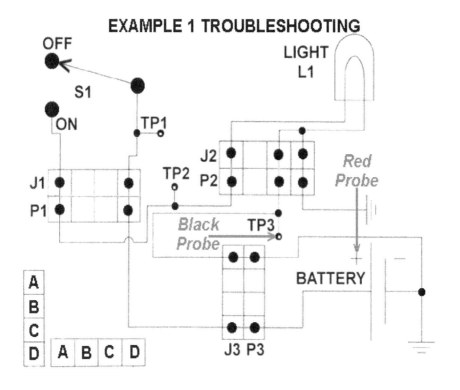

Assuming you have the correct 6VDC, use the – black probe to check for 6VDC at the ground or return line running thru line of J2/P2 pins C and D. See drawing on next page.

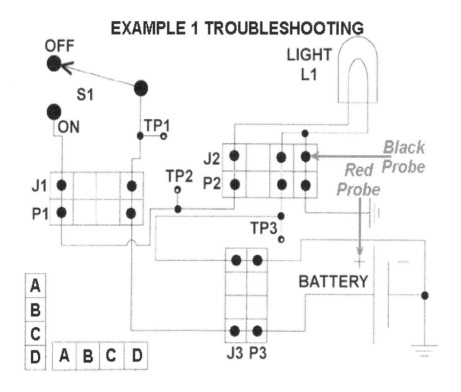

EXAMPLE 1 TROUBLESHOOTING

If both the positive or power side, and the negative or ground side check out OK, then the only thing left that could be bad is the bulb itself. This assumes you performed all the troubleshooting correctly.

In real life, you would change the light bulb before ever going thru all this troubleshooting. To check the filament, remove all power from the circuit and disconnect the battery from the circuit.

Measure for resistance and continuity from TP2 to TP3 using the multimeter. If you have continuity, then the filament is good. However, if you did everything else correctly, you will show no continuity and the filament is open.

The second method of completely using resistance or continuity checks also requires removal of all power from the circuit and disconnection of the battery from the circuit. Instead of measuring for voltage, measure for continuity using the same troubleshooting flow as in the Power On method. You are looking for minimum resistance between any two points.

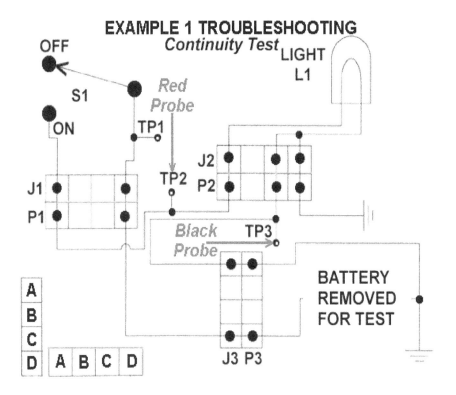

EXAMPLE 1 TROUBLESHOOTING
Continuity Test

Diagram Example 2

EXAMPLE 2

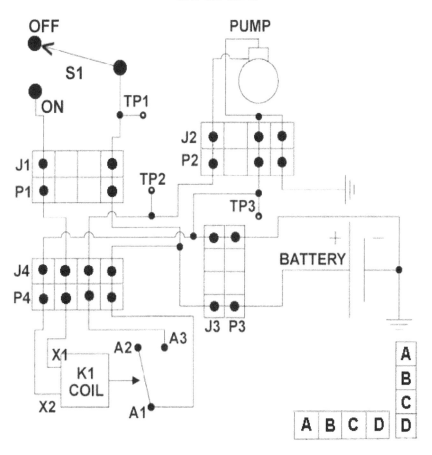

This is almost the same circuit as Example 1. The difference is that this one also has a relay and a pump (instead of the light) along with the 6-cell battery, connectors, SPST (single pole single throw) switch, and three test points.

The circuit operation follows: Positive voltage from the battery follows the wire thru connector J3/P3 pins D, then down the wire thru connector J1/P1 pins D past TP1 to the wiper common point of No (normally open) switch S1.

Again, normally open means that the switch is a break in the circuit. When the switch is "turned on" the circuit will be completed.

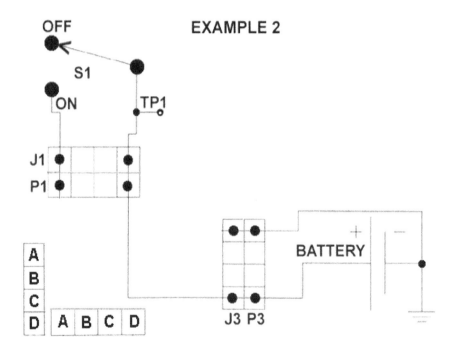

Remember, TP1 is a test point that allows the troubleshooter easy access to the circuit. It allows you to take measurements without disturbing the wiring or any protective circuit card coatings.

If S1 is closed, positive voltage follows the second wire thru J1/P1 pins A, thru connector J4/P4 pins B to the + side of the K1 coil at X1. The voltage then moves thru the coil (closing the Normally Open contacts) out K1X2 and obtains the ground return path thru J4/P4 pins A down the wires past a tie point thru J3/P3 pins A to the battery return and ground.

EXAMPLE 2

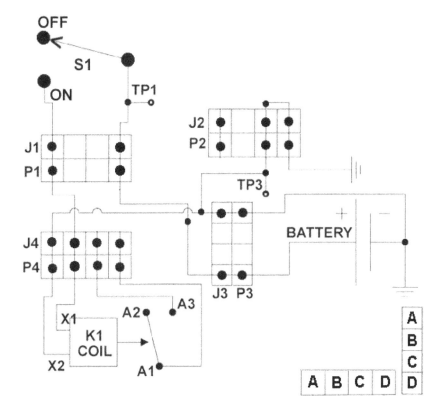

An alternate ground path exists from the common ground tie past TP3 thru J2/P2 pins C and back thru pins D to ground.

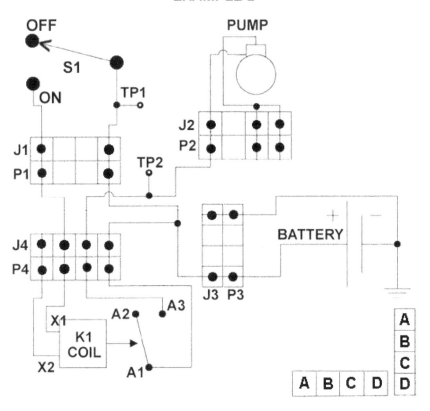

EXAMPLE 2

Let's back up a little to the common wire tie point just to the left of J3/P3 pins D. Remember that there is positive voltage on the wire attached to pin D. The tie point provides the positive voltage to J4/P4 pins D to K1A1 (relay common terminal). Note that the relay as shown is No or Normally open. If the relay is energized with voltage on the + side X1 of the relay coil, then contact arm will move up and close the circuit. This will allow the positive voltage on relay K1 terminal A1 to move thru A3 thru J4/P4 pins C, and up the wire past TP2 thru J2/P2 pins A to the pump motor. With Positive voltage present, the pump would operate if the circuit has a return ground path (and if the pump motor is good). Example on page 132.

EXAMPLE 2

PUMP

The return path is from the Negative side of the pump motor thru the negative lead thru a common tie point thru J2/P2 pins D and then to a chassis ground. An alternate path from the tie point is thru J2/P2 pins C past TP3 thru J3/P3 pins A to the battery negative and ground. As explained before, the return path for

K1X2 is thru J4/P4 pins A down the wires past a common tie point thru J3/P3 pins A to the battery return and ground.

How would you troubleshoot this circuit? Assume that the pump motor does not operate no matter what position switch S1 is in. There are two methods you can use, either with power on by measuring voltage, or by measuring continuity and resistance with power off and <u>disconnected</u> .

EXAMPLE 2 TROUBLESHOOTING

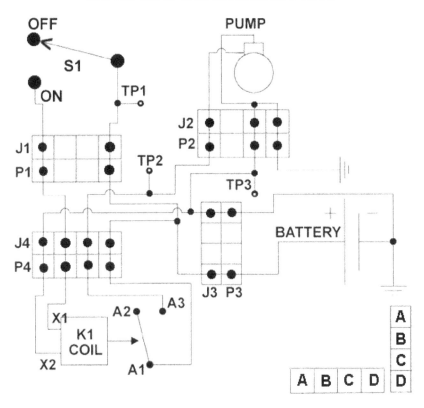

Power On - Using a multimeter, (after verifying correct operation of meter), establish that the battery has a charge by measuring voltage from the + terminal to the − terminal. Assume for this explanation that you get 6VDC. Keeping your negative black probe on the − battery terminal, touch your positive red probe to TP1. The meter should show 6VDC. If it does not, then there is a circuit disconnect in the wiring or connectors between TP1 and the + battery terminal. Disconnect J1/P1 and

EXAMPLE 2 TROUBLESHOOTING

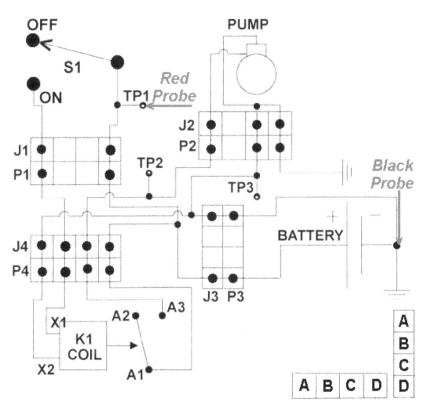

probe pins D for the 6VDC, and if not present, disconnect J3/P3 and look for the 6VDC at pins D. This should show you where the break is.

Assuming you have 6VDC at TP1, move your + probe to TP2 and check for voltage. You should have 0 VDC. Close (turn on) switch S1 and you will have 6 VDC at J4/P4 pins B. If not, the problem is either switch S1, J1/P1 pins A or the wiring to J4/P4 pin B.

EXAMPLE 2 TROUBLESHOOTING

To verify that Relay K1 is getting positive voltage to its common, touch the red probe to the common terminal K1A1. You should read 6VDC. If not, the problem could be at J4/P4 pins D or the common wire tie point after J3/P3 pins D.

EXAMPLE 2 TROUBLESHOOTING

Assuming that Relay K1 is operating correctly, check at TP2. You should have 6VDC. If so, you can check for 6VDC at J2/P2 pins A. If you have 0 at TP2, go back to J4/P4 pins C and see if the positive voltage is at pin C. If so, it is a wiring break between pin C and TP2. Example on next page (138).

EXAMPLE 2 TROUBLESHOOTING

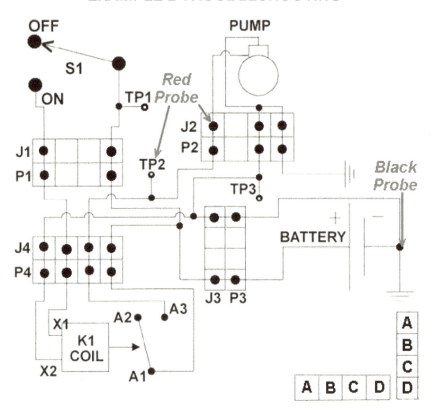

OK, assume that the power or positive side checked out fine. Let's look at the negative or ground side of the circuit. Everything after the pump motor is the ground side. To measure that using power on, you need to switch the position of the test probes.

Touch the + red probe to the positive terminal of the battery. Touch the – black probe to the two chassis grounds and TP3. The multimeter should read should indicate 6VDC. If not, look for breaks in the wiring or disconnect J3/P3 and probe pins A for voltage.

EXAMPLE 2 TROUBLESHOOTING

Assuming you have the correct 6VDC, use the – black probe to check for 6VDC at J2/P2 pins C and D. If you have voltage there, check for 6VDC at the negative input of the pump motor.

The last part of the ground return path that you need to check is the ground circuit for Relay K1.

EXAMPLE 2 TROUBLESHOOTING

Touch the black probe to J4/P4 pins A and verify that you read 6VDC. If not, there is a wiring break between J4/P4 pins A and TP3 or the chassis ground.

If both the positive or power side, and the negative or ground side check out OK, then the only thing left that could be bad is the pump motor itself. This assumes you performed all the troubleshooting correctly.

To check the pump motor, remove power from the circuit and disconnect the battery from the circuit. Measure for resistance and continuity from TP2 to TP3 using the multimeter. If you have continuity, then the motor is probably good.

EXAMPLE 2 TROUBLESHOOTING

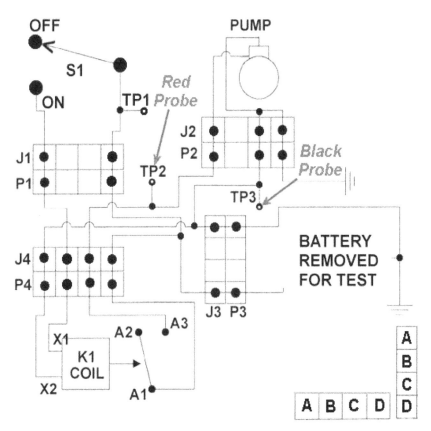

There are other things that could keep the motor from working, but for this exercise assume that the continuity of the pump motor is the deciding factor.

The second method of completely using resistance or continuity checks also requires removal of all power from the circuit and disconnection of the battery from the circuit.

Instead of measuring for voltage, measure for continuity using the same troubleshooting flow as in the Power On method. You are looking for minimum resistance between any two points. Remember, disconnect the power source or remove the battery from the circuit *__before__* doing resistance checks.

Diagram Example 3

Example 3 has more circuitry added, but is almost the same circuit as in examples 1 and 2.

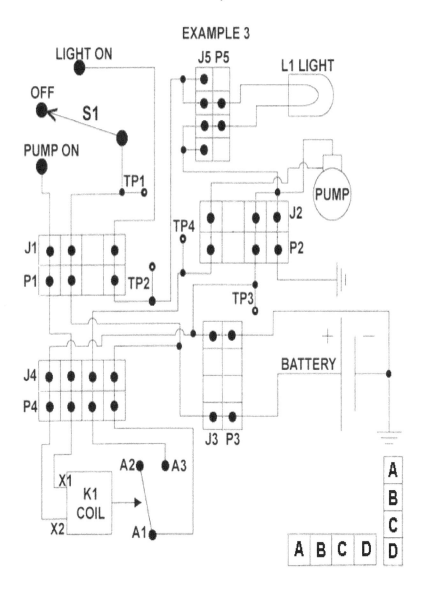

This third example also has a relay and a pump and an added light along with the 6-cell battery, five connectors, a SPDT (single pole double throw) switch, and 4 test points. It also adds three circuit breakers.

I will start where the circuit differs from Example 2. S1 is now a center-off SPDT. This extra position adds the LIGHT ON position. Assuming S2 is the LIGHT ON position,
positive voltage will flow thru J1/P1 pins D past TP2 thru J5/P5 pins B to the input lead off L1 LIGHT causing the L1 filament to glow.

The return path is out the other L1 lead thru J5/P5 pins C and J2/P2 pins D to the ground. An alternate ground path leads from J5/P5 pins C thru J2/P2 pins C past TP3 thru J3/P3 pins A to the Battery negative and ground.

Diagram Example 4

Example 4 has even more circuitry added, but is still based on the previous examples.

EXAMPLE 4

This fourth example also has two relays, a pump, and a light along with the 6-cell battery, six connectors, a SPDT (single pole double throw) switch, and four test points.

Since this circuit only adds another path and the circuit breakers to example 3, I will start where the circuits differ.

The battery feeds positive voltage thru J3/P3 pins D thru J6/P6 pins B, C, and D to the 3 circuit breakers. Pins A also receive power, and connect to TP4.

When Circuit breaker 1 (CB1) is closed it provides positive voltage thru J1/P1 pins B past TP1 to the S1 common terminal. When CB2 is closed, it provides positive voltage thru J4/P4 pins D to the relay common terminal of K1A1. When CB3 is closed, it provides positive voltage to the relay K2A1 common terminal.

S1 is still a center-off SPDT. This extra position adds the LIGHT ON position. Assuming S2 is the LIGHT ON position, positive voltage will flow thru J1/P1 pins D past TP2 thru J5/P5 pins B to the Relay K2X1 coil input. The flow will activate the coil and connect contacts K2A1 to A3. The return path for the K2 coil is out X2 thru J5/P5 pins A thru J2/P2 pins C to the ground. An alternate path leads J2/P2 pins C past TP3 thru J3/P3pins A to the Battery negative and ground.

When the K2 coil activates and connects contacts K2A1 to A3 (and if CB3 is closed), positive voltage will flow thru J5/P5 pins C to the input lead off L1 LIGHT causing the L1 filament to glow. The return path is out the other L1 lead thru J5/P5 pins D and J2/P2 pins C to the ground. An alternate path leads from J5/P5 pins C and J2/P2 pins C past TP3 thru J3/P3 pins A to the Battery negative and ground.

Troubleshooting of the additional circuits in examples 3 and 4 is performed using the same methods as in the circuits in examples 1 and 2.

How have you been doing? Are you ready for more advanced circuits? If the previous examples are not clear in your mind, go back and review the functional and troubleshooting explanations.

The next two examples are diagrams of commercial items from the years past. They are more difficult than the first four examples.

Diagram Example 5

Example 5 covers two pages. You will find that this split-wiring diagram format occurs often in large or complicated systems on autos, trucks, and aircraft.

In this exercise, you will have to line up the circuit paths to make sure you are on the correct one. You may find this easier if you photocopy the two pages and make one large wiring diagram for practice.

When working with multi-sheet diagrams, pay close attention when moving between pages or sheets so that you do not accidentally start following the wrong circuit. This sounds easy, but working with a 20 sheet wiring diagram is anything but easy.

In example 5, the fault is: Left Rear Turn signal does not light. Bulb and socket are good. Left Front Turn signal lights.

The full wiring diagram is on pages 150-151. The separate turn signal circuit path with troubleshooting hints and my troubleshooting is on pages 152 -153.

Try to come up with your own troubleshooting plan and problem solution before reading my solution.

EXAMPLE 5

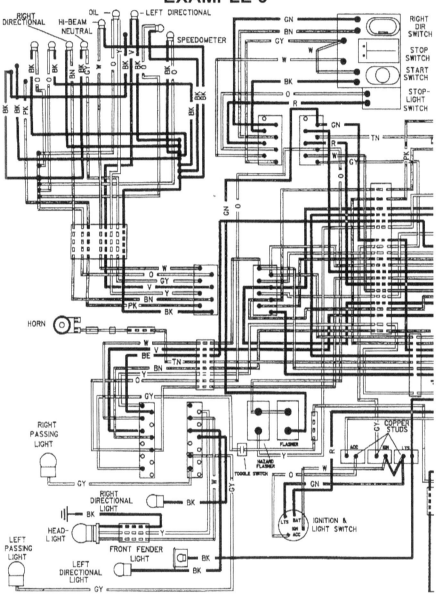

150

EXAMPLE 5

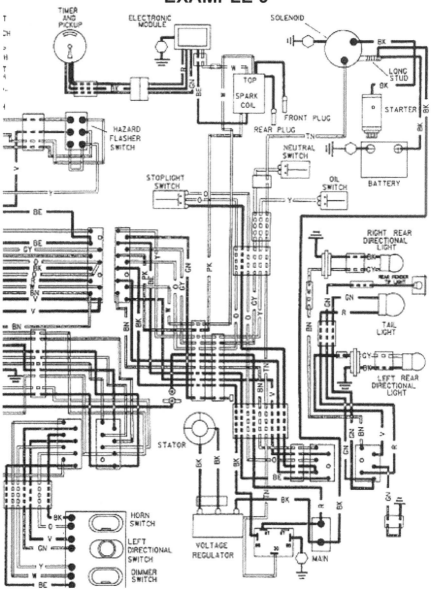

EXAMPLE 5

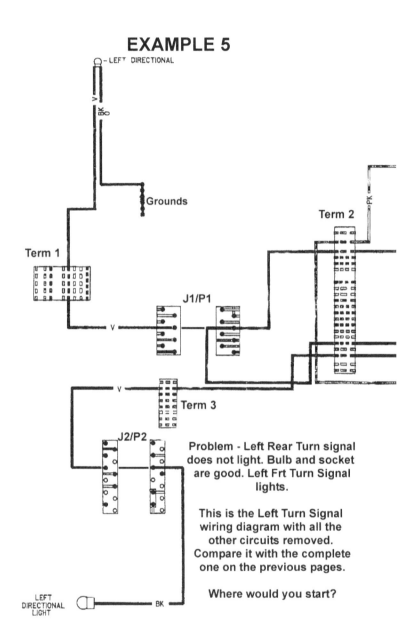

- LEFT DIRECTIONAL

V

BK

Grounds

Term 1

Term 2

PK

J1/P1

V

V

Term 3

J2/P2

Problem - Left Rear Turn signal does not light. Bulb and socket are good. Left Frt Turn Signal lights.

This is the Left Turn Signal wiring diagram with all the other circuits removed. Compare it with the complete one on the previous pages.

Where would you start?

LEFT DIRECTIONAL LIGHT

BK

EXAMPLE 5

What to do? Come up with your own plan before reading farther. See if we agree.

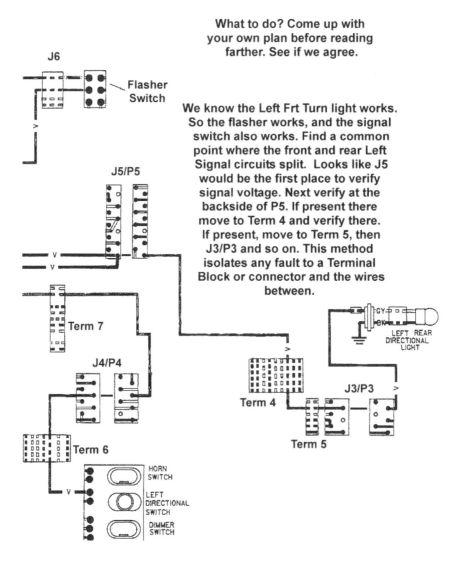

We know the Left Frt Turn light works. So the flasher works, and the signal switch also works. Find a common point where the front and rear Left Signal circuits split. Looks like J5 would be the first place to verify signal voltage. Next verify at the backside of P5. If present there move to Term 4 and verify there. If present, move to Term 5, then J3/P3 and so on. This method isolates any fault to a Terminal Block or connector and the wires between.

J6

Flasher Switch

J5/P5

Term 7

J4/P4

Term 6

HORN SWITCH

LEFT DIRECTIONAL SWITCH

DIMMER SWITCH

Term 4

J3/P3

Term 5

GY

BK

LEFT REAR DIRECTIONAL LIGHT

Diagram Example 6

Example 6 uses the same wiring diagram as Example 5 (on pages 150 - 151).

The fault you are troubleshooting is no 12VDC to the Spark coil. No other information is given. Assume nothing. Look at the full schematic and come up with a plan of attack.

Try to identify the circuits needing testing, the test points, and testing order. Highlighting the circuit path with a color pen will make it stand out from all the other circuits.

Sometimes, drawing out the circuit on another piece of paper will work best. The high tech computer way is to scan the page and then eliminate all the other circuits using your imaging software.

If you are having problems identifying the required circuits, look at the next page and compare it with what you have.

My approach to this problem and explanation of my troubleshooting sequence begins on page 156, following the wiring diagram. After you have come up with your plan, compare it with mine.

EXAMPLE 6

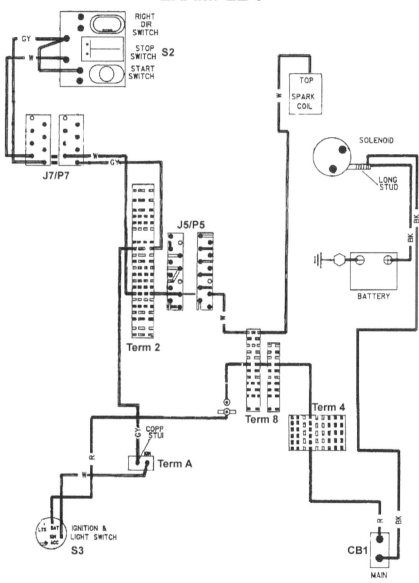

155

What to do? The problem is no 12VDC to the Spark coil. No other information is given.

The first thing to do is to verify Battery Voltage. This could be done a number of ways. You could grab your multimeter, and go right to the source and check for 12VDC across the Battery terminals.

If voltage was present, you could then work your way verifying 12VDC down the circuit all the way to the Ignition and Light switch. That would be perfectly correct.

I would probably short-cut that approach, and just turn the Ignition and Light Switch ON and see if any thing lit up.

If things lit up, then you know that the circuit to the Battery is good and the Battery is good. You would also know that the problem is between the Ignition and Light Switch and the Spark Coil

However, if nothing lit up, then the problem is between the Ignition and Light Switch and the Battery.

I would then go and grab the multimeter, and go right to the source and check for 12VDC across the Battery terminals. If voltage were present, I would work my way verifying 12VDC down the circuit all the way to the Ignition and Light switch.

156

EXAMPLE 6

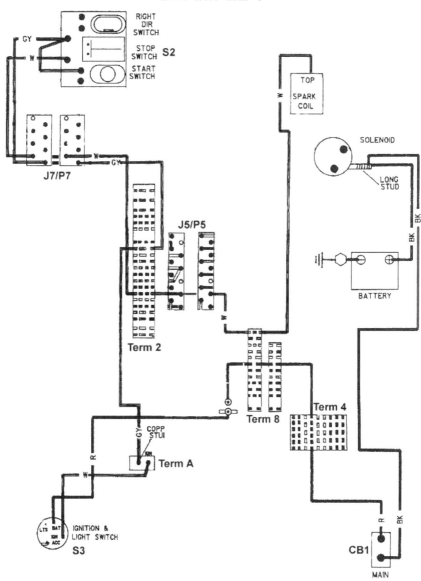

157

Let's split this example into two problems. In Problem A, indicators did light up when the Ignition and Light Switch was turned ON. In Problem B, indicators did not light up when the Ignition and Light Switch was turned ON.

Problem A – What do you know?
- You know that there is no 12VDC at the Spark Coil.
- You know that the circuit to the Battery is good.
- You know that 12 VDC Battery is good.
- You know that the problem is between the Ignition and Light Switch and the Spark Coil

What's your next move? I would use the multimeter to check for 12VDC at the following points in the following order:
- Term A Wire GY
- Term 2 Output side

While I am at Term 2, I would short-cut the troubleshooting and also check the other circuit going thru Term 2 (white wire) on its way to Connector J5/P5.

If there is 12VDC present, then the problem is on the J5/P5 circuit, and there is nothing to gain by continuing to follow the circuit thru Connector J7/P7.

EXAMPLE 6

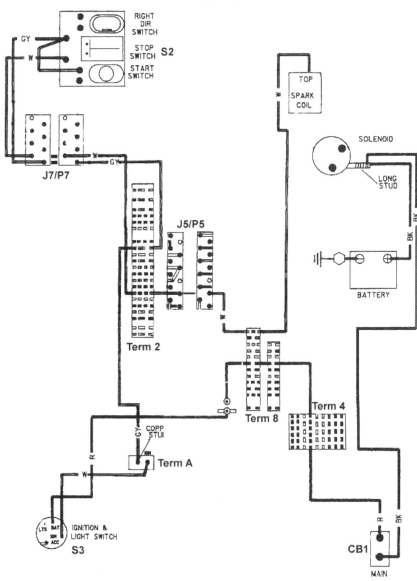

159

You have two choices:

- 12VDC at Term 2 on way to J7/P7 with No 12 VDC at Term 2 on way to J5/P5,
- 12VDC at Term 2 on way to J7/P7 with 12VDC at Term 2 on way to J5/P5.

First choice - If 12VDC at Term 2 on way to J7/P7 with No 12VDC at Term 2 on way to J5/P5, go to:

- J7/P7 Output side to Start/Stop switches
- J7/P7 Output side White wire

If J7/P7 Output side to Start/Stop switches has 12 VDC present, but J7/P7 Output side White wire does not have 12VDC, then you have isolated the problem to the wiring to and from the Start/stop Switch or the switch itself. Checking at the switch will give you your "Final Answer."

If both J7/ P7 Output side to Start/Stop switches and J7/P7 Output side White wire has 12VDC present, then the problem is the White wire from J7/P7 to Term 2 and that is your "Final Answer."

Second choice - If 12VDC at Term 2 on way to J7/P7 with 12VDC at Term 2 on way to J5/P5, there is nothing to gain by continuing to follow the circuit thru Connector J7/P7. I would verify 12VDC at :

- J5/P5 Output White wire
- Term 8 White wire

EXAMPLE 6

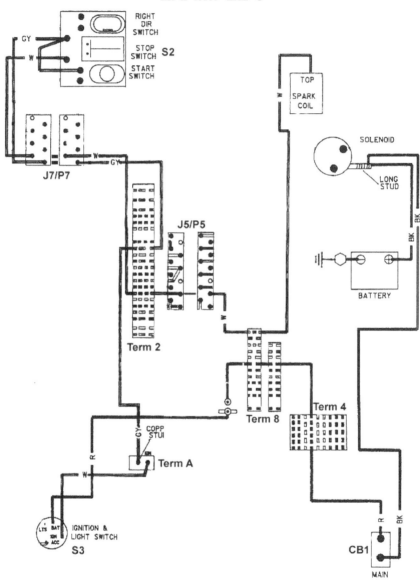

161

If you verify 12VDC at the Term 8 Output white wire, but still NO 12VDC at the Spark Coil, then the problem is the White wire between Term 8 and the Spark Coil. But, what if when you checked at the Spark Coil there was 12VDC present there, and you had not repaired anything.

Again two choices:
1) 12VDC was at the Spark Coil all the time. (You know that this was not possible because you verified that the 12VDC was not at the Spark Coil at the start of the troubleshooting).

2) You accidentally fixed the problem as you were troubleshooting the circuit. This happens more often than you would think. It may have been a deformed connector pin, loose connector, oxidation or corrosion, or something you accidentally disturbed with our voltage testing.

In any case, it is now fixed and you are the "Hero." Note that the problem may come back and visit you again, but that is no sweat because now you "know the rest of the story."

EXAMPLE 6

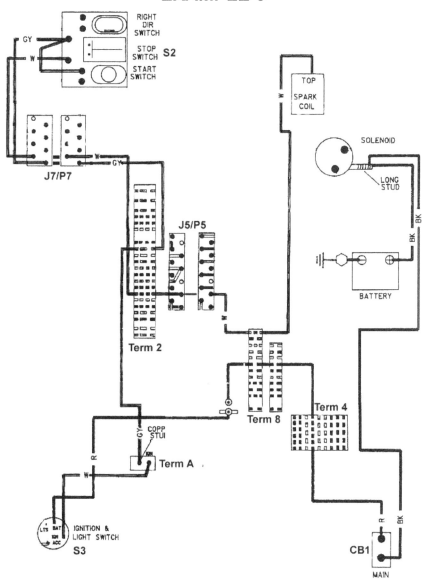

163

Practical Exercises

I wrote the first edition of this book for one of the University classes that I teach. By this time in the course, the class would have covered safety, basic theory of troubleshooting, and studied the troubleshooting exercises, just as you have in this book.

I found that what the class needed at this point was more time and experience "contemplating" electrical circuits. The class would move into what I called "free thinking team exercises." The students would form teams and would randomly draw the name of an electrical appliance from a fish bowl.

The team task was to design and sketch the electrical circuit for the appliance they drew. When they finished, they would put up the design for the rest of the class to see and review. This method gave them the circuit experience that they needed.

So what is so important about this experience? To be a successful electrical troubleshooter, you need to be able to look at an electrical appliance or machine and be able to visualize in your mind the electrical circuitry and operation before you ever refer to the wiring diagram.

Does this mean I am recommending not using the correct wiring diagram? NO! It means that if you already can see the circuitry in your mind, then it will be a whole lot simpler to understand the wiring diagram when you do pull it out.

This ability is a skill, and practice is the path to this skill. You cannot just read about it in this book, and acquire the ability. While this method was designed for a Team approach, I have modified it for an individual approach here.

Pick one of the appliances from the following list. Using the examples and information in the preceding pages, draw "your best" idea of what the electrical circuit should look like.

After you are finished, go to the following section – **EXERCISE CIRCUITS** - and find my wiring diagram for the appliance you selected. Compare the two, and hopefully they should be similar.

Look for any major differences. If there are some, think about why that is. Follow the circuit path on my design, and then compare it to your circuit path. Modify your design as needed. Eventually, both your design and mine should be close.

Note that there are different ways of achieving the same result, so even though your design does not match mine, it may still work.

Electrical Component List

Battery Flashlight
120VAC to 10VAC Wall Transformer
16VAC House Door Bell
Battery Pencil Sharpener
120VAC Electric Charcoal Lighter
120VAC Portable House Fan
120VAC to 12VDC Electric Train Set
24VDC Electric Golf Cart
Attack Helicopter Lighting Circuit

How should you start? OK, you have selected your appliance.

First, think about what the appliance accomplishes, i.e., a vacuum sucks air.

Second, what does the appliance need to accomplish what it does, i.e. a vacuum needs a motor.

Third, how do you control the motor, i.e. a switch.

Fourth, how and where does it get its power, i.e. batteries or wall 120VAC.

Fifth, do not forget the ground or return line. Everyone remembers the Hot or power line, but the ground is equally important.

Other things to think about are relays (for high current), temperature sensors and thermostats, fuse and circuit breakers, electrical plugs and connectors, grounds, wires, fusible links, variable controls, etc.

Make a rough block diagram showing everything you think you need. Use color pens or markers to draw your wires or circuits connecting the blocks. Follow each circuit path and see if it does what you want. If it does, redraw it clearly and check it again.

Some basics are: All circuits need protection, i.e. fuses. A circuit that draws 1 amp will not be protected by a 10-amp circuit breaker or fuse. Relays are used for high amperage items, and most motors will require the use of a relay. Relays need power and grounds to their coils in order to operate. All circuits must have a LOAD…otherwise they are actually a "short-circuit."

Satisfied with your design? Time to compare with mine in the next section. I suggest you do all the appliances for experience.

See if you can improve on my designs. If you can, then you have a good understanding of how electrical circuits work.

Exercise Circuits

IMPORTANT SAFETY NOTE/WARNING:
These wiring diagrams are for educational purposes only. I do not propose or suggest that there is enough information in any of these diagrams to allow you to safely construct or operate these items. The information is general and does not have necessary safety information included. <u>Do not build any of these items from the following information.</u> Again, these diagrams are for educational purposes only.

Battery Flashlight Circuit

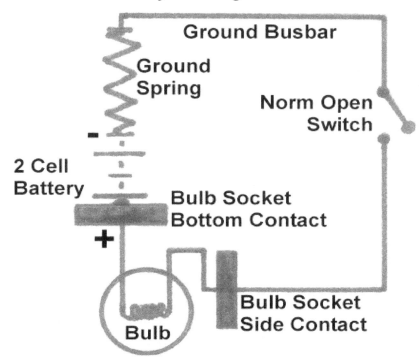

Is your drawing basically the same? If not, evaluate any differences to make sure you understand the theory. Note that on this circuit the switch connects and disconnects the ground or negative side. Compare this with the list of possible faults starting back on page 65.

120VAC to 10VAC Wall Transformer Circuit

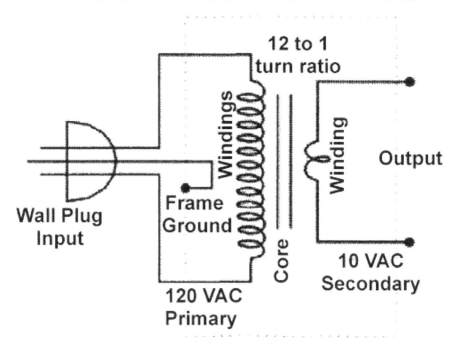

How does your drawing compare? Note that on the 120VAC primary side there are 12 turns or windings, while on the 10VAC secondary side there is only 1 turn or winding... In reality, there would be many more turns on both sides, but the 12:1 ratio would still hold. This gives us lowered voltage and

170

increased amperage. Varying the turn ratio will vary the secondary AC voltage output and current. Since there is no rectification shown, this will give us AC not DC.

16VAC House Door Bell Circuit

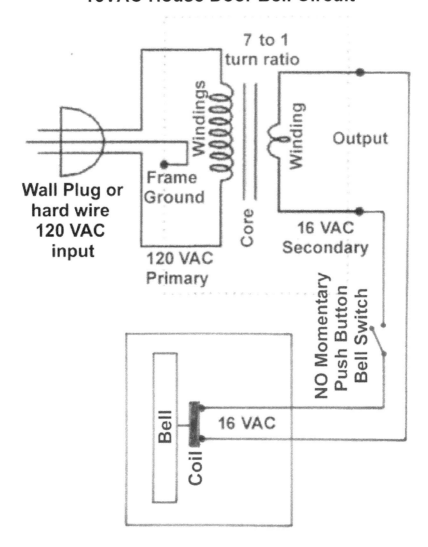

Is your drawing basically the same? If not, evaluate any differences to make sure you understand the theory. Note that on the 120VAC primary side there are now 7 turns or windings, while on the 10VAC secondary side there is only 1 turn or winding... In reality, there would be many more turns on both sides, but the 7:1 ratio would still give the 16VAC secondary output for the bell coil. The only other components are the NO (normally open) momentary push button bell switch and the 16 VAC coil in the bell assembly.

Battery Pencil Sharpener

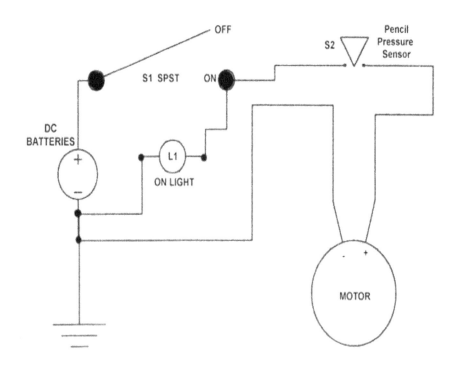

How does your drawing compare? If not, evaluate any differences to make sure you understand the theory. The pencil sharpener has a light (L1) to indicate that the power switch (S1) is on. S2 is pressure switch (safety feature) that is activated by the pencil being inserted in to the Sharpener hole. The motor will not power up unless S2 is depressed.

120VAC Electric Charcoal Lighter

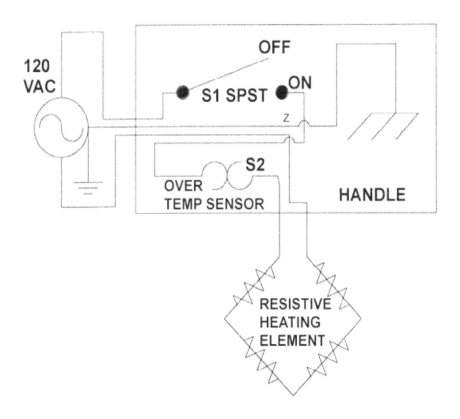

The 120 VAC Electric Charcoal Lighter shares a common feature with all electric ranges, ovens,

griddles, etc. It uses a resistive element that gets very hot, but not hot enough to melt. Eventually, the elements do burn out and require replacement.

The resistive wire is usually nichrome wire encased in ceramic insulation with a steal covering over the ceramic. S1 is a standard single pole single throw slide switch. S2 is bimetal switch to prevent the unit from overheating and destroying itself. The bimetal switch contains two dissimilar metals joined together. They form one unit with a differential expansion rating making it bend and disconnect if it reaches an over temperature. Note that the frame handle is grounded back to the 120VAC source.

Is your drawing basically the same? If not, evaluate any differences to make sure you understand the theory.

120VAC Portable House Fan

The Portable House Fan has single pole single throw switch (S1) for on and off. No power light is included since the fan movement will indicate whether power is applied. The speed of the fan is controlled by stepper switch S2 that moves thru 4 speeds from low to high. These circuits connect to individual windings in the fan motor that control the motor speed.

Is your drawing basically the same? If not, evaluate any differences to make sure you understand them.

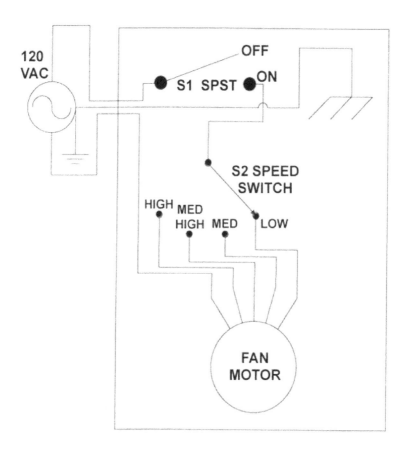

120VAC to 12VDC Electric Train Set

This diagram is based on a three-rail "O" scale train set similar to the Lionel sets. The three rail design uses the two outer rails as the negative return and the center rail as the positive supply. The locomotive outer drive wheels roll on the outer rails to connect with the negative return. A center pick-up beneath the locomotive rides on the center rail and transfers the positive supply voltage to the motor.

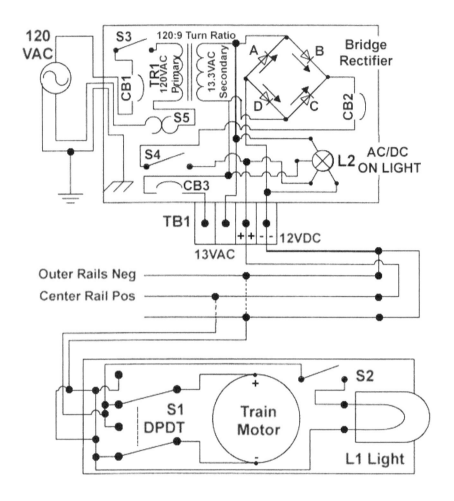

The 120VAC primary transformer has a turn of 120 to 9, which gives a secondary voltage of around 13.3 VAC. The 13.3 VAC is tapped off and run to TB1. This is for use of train set accessories such as lights, sirens, etc. A full-wave bridge rectifier is also wired to the transformer secondary.

Making up the bridge rectifier are four diodes (labeled A,B,C, and D). The circuit operates as follows: In the forward half-cycle "B" conducts with the diode (follow the anode arrow) out to the load (train motor) and returns back thru "D" (with the arrow). In the reverse half-cycle, "A" conducts with the diode (follow the anode arrow) out to the load (train motor) and returns back thru "C" (with the arrow). Remember that this is happening at 60 times per second.

The bridge rectifier does not give out "pure" straight-line DC. Its output is still a sinusoidal wave as is the primary AC. Each diode creates a 0.6V voltage drop as it conducts for a total of 1.2V per side. The bridge rectifier output to the train motor about 12.1 VDC. The DC voltage formula is as follows: Primary 120VAC/9 turns =13.3VAC secondary – 1.2V voltage drop = 12.1VDC bridge output.

CB1 protects the transformer while S3 is the main ON-OFF. CB2 protects the bridge rectifier, and CB3 protects the transformer secondary. S5 is a temperature overheat sensor that protects the transformer. S4 controls the 12VDC output to TB1. L2 is the "Power On" light and indicates the presence of both the 13.3VAC and the 12.1VDC.

In the train locomotive, S1 is a double pole double throw (dpdt) switch serving as the forward and reverse switch. It changes the polarity seen by the motor causing forward or reverse movement. S2 is a single pole single throw (spst) switch turning on or off the locomotive headlamp.

Is your drawing basically the same? If not, evaluate any differences to make sure you understand them.

24VDC Electric Golf Cart

The 24VDC Electric Golf Cart uses two 12VDC batteries in series to provide 24VDC for the drive motor. This means that you have two separate power circuits, 1) 24VDC and 2) 12VDC. The 12VDC circuit powers all the lights, horn, relays, etc. This allows the use of economical 12VDC auto electrical parts in the cart. One ground circuit works for both power circuits.

The 24VDC circuit functions as follows: Positive 24VDC is present at CB1. Closing CB1 puts power at the Relay K1A1. CB2 of the 12VDC circuit is now closed. This puts +12VDC at the fixed terminal of S1 Main Power. Closing S1 lights Power Lamp L1 and energizes Relay K1 coil at thru X1. This closes the K1A1-A2 and B1-B2 contacts. The closing of the A2 contacts puts +24VDC at the Relay K2A1 contact. Relay K2 activation is controlled by the S4 seat pressure switch. S4 is powered by the +12VDC circuit from the K1B1-B2 contacts thru TB1 thru F5 fuse. This is a safety feature and requires someone to be sitting in the seat for the golf cart to move.

Someone is sitting in the seat, so S4 is closed and Relay K2 is powered thru X1. This feeds +24VDC to S7 Speed selector switch (the accelerator). There are 4 speeds available – slow (R1), medium (R2), med high (R3), and high (R4) depending on which resistor circuit is selected. The resistors are of different ohm values with R1 having the most resistance and R4 being a direct wire (minimal resistance).

Follow the circuit as if S7 is connecting with R4 (H or High). This puts +24VDC at TB3 to feed thru Diode CR4 to the common tie point. The diodes prevent circuit back-flow. All four resistive–diode circuits work the same though with different voltage outputs. The circuit continues with +24VDC at Forward-Reverse Switch S9 A2 and B3. This DPDT switch provides for reversing the polarity to the drive motor therefore reversing the direction of rotation and travel. The S9 A2-A1 path provides the normal forward motion with +24VDC provided to the + Drive Motor terminal. At this same time, the S9 B2-B2 circuit connects with the – Drive Motor terminal to provide the required return circuit thru TB4 back to the battery negative source.

For reverse, the +24VDC is routed S9 B3-B1 and is put on the – Drive Motor terminal while the S9 A3-A1 connects the battery negative to the + Drive Motor terminal. Note that to make this all work, the Drive

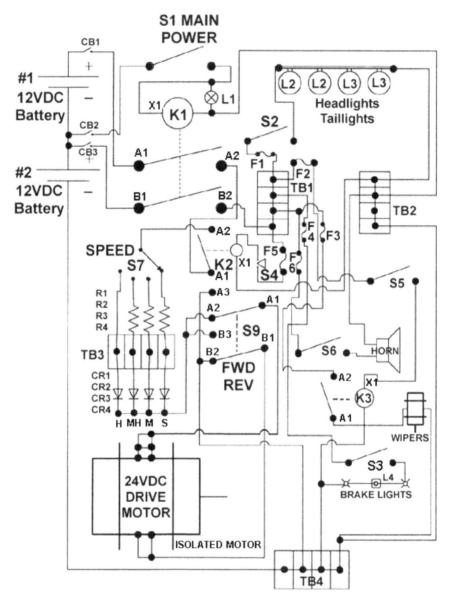

Motor is insulated and isolated from any electrical ground. Optionally, you could have a third set of contacts in S9 switching the applicable motor ground.

This completes the 24VDC Drive Motor circuit.

The +12VDC circuit provides for activation of the +24VDC Drive Motor circuit (explained above). It also provides power for the horn (TB1 thru F6 fuse thru S6 to horn), the windshield wipers (TB1 thru F2 thru S5 to Relay K3 X1 and coil). K3 close the K3 A1-A2 contacts providing +12VDC from TB1 thru F3 to the wiper motor. The headlights and taillights (L2 and L3) are powered from TB1 thru F1 thru S2. The brake light is powered from TB1 thru F4 thru Brake pedal switch S3 to the L4 brake lights. All grounds for the 12VDC system are thru either TB2 or TB4.

Is your drawing basically the same? If not, evaluate any differences to make sure you understand them.

Attack Helicopter Lighting Circuit

This may be disappointing for you, but the circuit is very simple. It consists of a power source (115VAC 400HZ), numerous connectors and wires, a circuit breaker, a variable isolation transformer, and the lights.

Circuit operation is as follows: 115VAC is provided by one of the aircraft AC phases (there are three – A,B,C) thru various wires and connectors to the Utility AC Bus to the CB1. There is no On-Off switch as the CB1 also performs that function. More wires and connectors feed the 115VAC to the primary of

TR1. TR1 is both a step-down 1:12 isolation transformer (keeping the two circuits electrically separate), and a variable voltage controller. This allows for dimming the intensity of the lights. More wires and connectors feed the reduced AC power to the various lights.

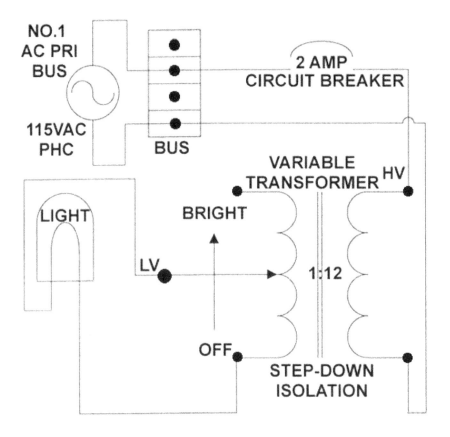

Is your drawing basically the same? If not, evaluate any differences to make sure you understand them.

Comprehension Questions

This section is designed to help you evaluate how much of the information that you have read in this book has been retained, or actually understood. Try to pick the BEST answer. My answers are in the next section.

1._____ flows in one direction, reverses, and then flows in the opposite direction.
a. DC
b. Magnetized
c. Inductive
d. AC
e. None of above

2._____ means that electrons only flow in one direction.
a. AC
b. Magnetized
c. Inductive
d. DC
e. None of above

3._____ occurs when one wire that has no physical connection to the second wire, cause a current in the second wire.

a. Induction
b. Magnetism
c. Transformation
d. Capacitance
e. None of above

4. A _____ circuit has only one path for current flow.
a. Parallel
b. Series
c. Series-Parallel
d. Parallel-Series
e. None of above

5. A(n) _____ circuit that has more than one path or leg for current flow.
a. Parallel
b. Series
c. Intermittent
d. Electronic
e. None of above

6. The _____ leg refers to the section of the circuit that carries current to a load.
a. Ground
b. Negative
c. Hot
d. Parallel
e. None of above

7. One of the most basic rules in electricity is that _____ follows the path of least resistance.
a. Ohms
b. Wattage
c. Voltage
d. Current
e. None of above

8. "Normally open" means that there is no complete current path when the switch is in its _____ position.
a. On
b. Abnormal
c. Closed
d. Unswitched
e. None of above

9. The _____ side is the input side of a transformer.
a. Secondary
b. Primary
c. Base
d. Primary
e. None of above

10. What do you call it when two points with different voltage levels (i.e. hot and ground) are connected with no resistance or load between the two points?
a. Resistance
b. 28 VDC
c. Short
d. Wattage
e. None of above

11. Conductive articles of jewelry and clothing may be worn around exposed energized parts.
a. True
b. False
c. Maybe
d. Always

12. A(n) _____ will normally increase total circuit _____ and reduce total circuit current.
a. Fuse; voltage
b. Open; resistance
c. Short; resistance
d. Fuse; capacity
e. None of above

13. A(n) _____ causes more _____ and less resistance.
a. Open; current
b. Open; voltage
c. Short; current
d. Load; voltage
e. All of above
f. None of above

14. Which of the following is not something you should consider when forming a troubleshooting plan?
a. Prior occurrence
b. Recent modifications or changes
c. What is working
d. What is not working

e. All of above
f. None of above

15. Redundant circuits are second circuits that provide duplicate electrical paths. Why are they used?
a. Carry more current
b. Return path to ground
c. Increase battlefield survivability
d. Carry more voltage.
e. None of above

16. The influence or occurrence of "LRU Tolerances"
a. Makes effective troubleshooting easier
b. Relates to voltage spike resistance
c. Relates to LRU temperatures
d. None of above

17. What is a Mux Bus?
a. Transformer
b. Refers to inductive power transfer
c. Different signals over one circuit path at same time
d. None of above

18. "Induced failure" relates to the theory that two inductive components (i.e. transformers) can pass a virus from one to another.
a. True
b. False
c. Neither

19. The very first thing you should verify after turning an analog multimeter on.
a. Insulation is good on leads
b. Meter is set to 10A
c. Component to be tested power is on
d. Meter is set to 100A
e. Meter is Zeroed
f. None of above

20.   You are checking continuity and resistance between two truck electrical connectors 21 ft apart. You decide to use the truck frame and chassis as the return line for your multimeter resistance reading. What may be a down-side or problem with using this technique?
a. Meter battery discharges quicker
b. Leads are too long now
c. Added resistance distorting reading
d. There is no problem with technique
e. None of above

21. What is a better way of performing the continuity and resistance check in question #20.
a. Use two meters
b. Use bigger battery
c. Use external long wire for return
d. None of above

22. You have an <u>analog multimeter</u> to measure DC voltage across a device. You properly connect the

probes to the terminals on the device, and the meter needle indicates 0 volts. This could indicate what?
a. Multimeter is not on
b. Multimeter fuse is open
c. Multimeter is broken
d. 0 volts across device
e. Probe wire open
f. All of above
g. None of above

23. You have a <u>digital multimeter</u> to measure DC voltage across a device. You properly connect the probes to the terminals on the device, and the meter indicates 0.00 volts. This could indicate what?
a. Multimeter fuse is open
c. Multimeter is broken
d. 0 volts across device
e. Probe wire open
f. All of above
g. None of above

24. Which of the following is crucial to the correct operation of a Mux Bus?
a. Unique component address
b. Shared addresses
c. Large computers
d. US Post Office
e. None of above

25. If all other factors remain the same, what would be the effect of increasing from AWG (American Wire Gauge) 22 to AWG 26 in a wire run or circuit?

a. Wattage will decrease
b. Resistance will increase
c. Wattage will increase
d. Resistance will decrease

26. Why do you shield wiring?
a. Makes wires tougher
b. Increases resistance
c. Reduce capacitance
d. Reduce risk of outside EMF influence

27. What type of circuit is shown below?

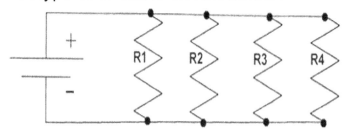

a. Resistive
b. Balanced
c. Series
d. Parallel
e. None of above

28. What type of circuit is shown below?

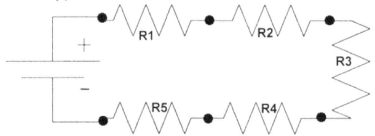

a. Series-Parallel
b. Resistive
c. Short
d. Series
e. Parallel
f. None of above

The next two questions (**29** and **30**) refer to the diagram below. This circuit is positive grounded, not negative grounded. All CBs (circuit breakers) and switches are currently OPEN. Selector switch S2 is in the position shown (connecting to 4). The battery is a fully charged at 6VDC.

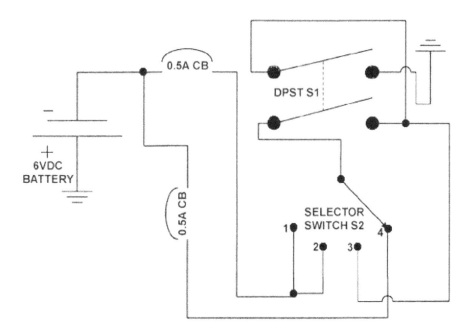

29. In the following order, you close both CBs and move S2 to position 2. What occurs when you close S2, and why?
a. Nothing, Circuit incomplete
b. Nothing, No connection to ground
c. Top CB immediately opens, Short to ground
d. Battery quickly discharges, Runaway current
e. None of above

30. You again open both CBs and S2. In the following order, you close both CBs and move S2 to position 3. You close S2. What occurs and why?
a. Top CB immediately opens, Short to ground
b. Nothing, No connection to ground
c. Lower CB immediately opens, Short to ground
d. Battery quickly discharges, Runaway current
e. None of above

31. One method of troubleshooting involves using substitution or exchange. What are the benefits of using this method?
a. Less risk
b. Cheaper
c. Sometimes quicker
d. Always fixes problem
e. All of above
f. None of above

32. Low voltage systems have more intermittent connector problems than high voltage systems.
a. True

b. False

c. Neither

33. Which of the following is a function of a diode?

a. Transformer

b. Storage

c. Amplifier

d. Rheostat

e. Rectifier

f. None of above

34. Which of the following are advantages of the Mux Bus?

a. Low cost

b. Versatility

c. Standard

d. Weight reduction

e. All of above

f. None of above

35. What is my suggested method of troubleshooting the Mux Bus?

a. Leave to specialist

b. Oscilloscope

c. Component exchange or substitution

d. None of above

36. Troubleshooting of wiring problems in a Mux Bus can be performed using a multimeter.

a. True

b. False

The remaining questions refer to the wiring diagram on pages 196 – 197. This diagram is typical of a one you might find on any job. This circuit does not do anything – it is for educational use only. These are tough questions with a number of variables and conditions provided. Some of the information is useful and some is useless, just like in the real world. Pay attention to all the details. There is a single page copy of the diagram for reproducing on page 203 if you want copies to practice on.

Read the questions carefully. Small facts will make significant differences. Some questions may require multiple choices.

37. Using a multimeter, you check for continuity on the circuit path by placing the + red lead on P3303 pin M and the – black lead on J3303 pin P. The meter shows low ohms resistance. When you check for continuity by placing the – black lead on K1X1 and the + red lead on J3303 pin P you show what appears to be infinity. What is the fault?
a. There is no fault
b. Multimeter failure
c. K1X1 is a grounded circuit
d. None of above

38. All CBs are open, TB3 pin 1 is now at 0 volts. K1 is in non-powered state. The light is off. S1 is in the normal position. Today is Monday. Friday will be the fifth day of the week. What is the resistance from TB3 pin 1 to chassis ground?

a. 170 watts
b. 0
c. 170 ohms
d. Infinity

39. CB4 is now closed. It is now Tuesday. The K1 coil now has 0 ohms resistance across K1X1 and K1X2. K1X2 has infinite resistance between it and TB2 term 2. What is the resistance from TB3 pin 1 to chassis ground?

a. 170 watts
b. 0
c. 170 ohms
d. Infinity

40. TB3 pin 1 has +28vdc present. You close CB3. You close CB2. K1 coil resistance of 170 ohms. This is question 29. K1X1 is at ground potential. The resistances across K1B2 to K1B3 and K1A2 to K1A3 are both 0.8 ohms. You close CB4. The continuity reading across K1A2 to K1A1 is 0.8 ohms while K1A1 to K1A3 is infinite. All wiring is connected as shown on diagram. What is the resistance from J3303 pin A to P3303 pin H?

a. 170 watts
b. 0
b. 170 ohms
c. Infinity

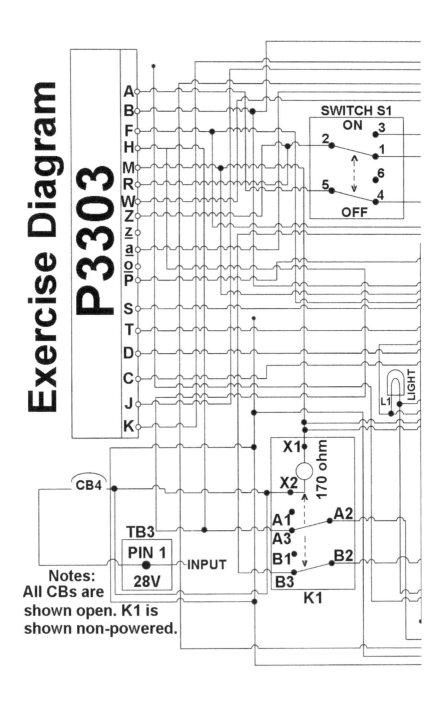

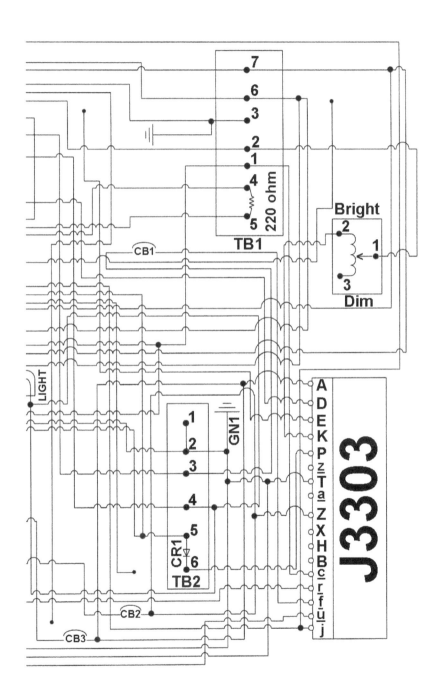

197

41. Refer to conditions set in Question #40. What is the voltage difference between the two CB4 terminals? What is the resistance between the two CB4 terminals?
a. 0 volts, 170 ohms
b. Infinity, Infinity
c. 0 volts, none
d. 28 volts, 0 ohms
e. None of above

42. The ground potential at L1 terminals 1 and 3 is infinite. P3303 pin Z is now at +28vdc potential. S1 is in "ON". R1 has a 220 ohm value. L1 is a correct voltage serviceable correctly installed light. The resistance between L1 term 3 and TB1 term 1 is 100 ohms. What would it take to make L1 light up? (Requires 2 answers).
a. More power, Scotty
b. S1 in OFF position
c. Won't ever light up
d. Ground potential at J3303 pin c
e. Ground potential P3303 pin R

43. Assume that this is one part of a military aircraft. CBs 1, 2, and 3 are closed. The wiring from J3303 pin A to CB3 is completely cut apart by fragments from an enemy round. When you run a resistance check there still is continuity between J3303 pin A and K1A2. A second resistance check shows J3303 pin A is grounded. With the information provided along with the diagram as drawn, what is one possible cause of this?

a. Normal circuit operation
b. K1A2 providing the ground
c. Damaged wire is contacting the aircraft frame
d. K1A3 providing the ground
e. None of above

44. There is 28v at TB3 pin 1. Only CB4 is closed. You hear the relay contacts move. There is continuity between K1 terminals B2 and B3. There is also continuity between K1 terminals A1 and A2. You open CB4. There is 0.8 ohms resistance between K1 terminals A2 and A3, but infinity from A2 to A1. There is continuity between K1 terminals B2 and B3. Ground GN1 has failed and is now open. You again close CB4. There is infinity between K1 B1 and B2. You also close CB2 and CB3. What is the resistance reading from J3303 pin Z to P3303 pin F?
a. Infinity
b. 0.8 ohms
c. 28 volts
d. 170 ohms
e. 220 ohms
f. None of above

45. Same conditions as Question #44. What effect does the failure of Ground GN1 have on wire shielding?
a. Better
b. Worse
c. No effect
d. Insufficient information

e. None of above

46. You are checking for continuity. You find 150 ohms resistance between J3303 pins c and f. Switch S1 is in the "On" position. There is continuity P3303 pin A J3303 pin D. There is continuity between P3303 pin Z and R, but infinity relative to any other pins in either P3303 or J3303. What is the most probable fault?
a. Lamp L1 burned out
b. TB2 pins 3 and 4 shorted
c. Switch S1 burned out
d. CB1 open
e. None of above

47. Refer to Question #46. Would your answer be different if Switch S1 was in the "Off" position?
a. Yes
b. No
c. Maybe
d. Insufficient information
e. None of above

Questions #48, #49, and #50 use the following conditions. (Careful... these are tricky).

You know, observe, or have measured the following:
• K1X1 to chassis ground resistance 170.7 ohms
• +28vdc at TB3 pin 1
• CB3 and CB4 closed
• CB1 and CB2 are open

- +28vdc at K1X2
- GN1 is a ground point
- K1B3 to K1B2 resistance infinite
- Continuity from K1B1 to K1B2
- Infinite resistance K1B3 to K1B1
- Resistance K1A3 to K1A2 approx 0.7 ohms
- CR1 is open and faulty
- +28vdc at J3303 pin A
- P3303 pin R and P3303 pin H are electrically jumpered together (connecting wires)
- S1 is in ON position
- P3303 pin Z and P3303 pin W are electrically jumpered together (connecting wires)
- J3303 pin K and J3303 pin f̲ are electrically jumpered together (connecting wires)
- TB2 pins 3 and 4 are electrically jumpered together (connecting wires)
- TB1 pins 1 and 3 are electrically jumpered together (connecting wires)
- L1 is lighted

48. What effect does the dimmer switch have on L1?
a. L1 Light dims
b. L1 Light immediately goes out
c. No effect on L1
d. What L1 light?
e. None of above

49. What effect does the on or off position of Switch S1 have on L1?
a. Both turn it off

b. No effect

c. None of above

50. What components have failed? (Two answers)

a. L1 is defective

b. S1 intermittent

c. K1A2 & A3 are shorted or stuck together

d. CR1 open

e. None of above

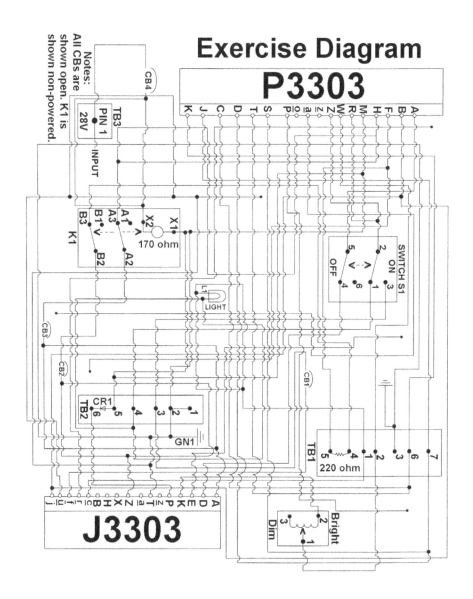

Exercise Diagram for photocopying

Comprehension Answers

The correct answer is underlined in bold italics. A Reference page that generally explains the answer is also listed. Some answers also have a more detailed explanation.

Note on the Wiring Diagram Questions: I teach and troubleshoot from the positive source to the ground or negative side. That is the way I was taught and what works best for me. Others teach and troubleshoot from the ground or negative side to the positive source. It does not really matter which way you do it. Experiment and find what works best for you. However, the explanations you find here are written using positive source to ground side.

1. ***AC*** flows in one direction, reverses, and then flows in the opposite direction. Answer: ***d. AC***
Reference: Page 3.

2. ***DC*** means that electrons only flow in one direction. Answer: ***d. DC***
Reference: Page 3.

3. ***Induction*** occurs when one wire that has no physical connection to the second wire, cause a current in the second wire. Answer: ***a. Induction***
Reference: Page 39.
Explanation: See image and text below.

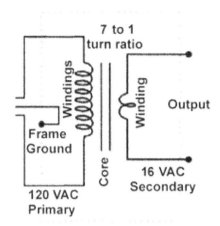

7 to 1
turn ratio

Windings

Winding

Output

Frame
Ground

Core

120 VAC
Primary

16 VAC
Secondary

Induction works on the principle that when there is current in one wire, induction can cause current in a nearby second wire that has no physical connection to the first. The most common inductor is the transformer.

4. A **_Series_** circuit has only one path for current flow.
Answer: **_b. Series_**
Reference: Page 4.

5. A(n) **_Parallel_** circuit that has more than one path or leg for current flow. Answer: **_a. Parallel_**
Reference: Page 5.

6. The **_Hot_** leg refers to the section of the circuit that carries current to a load. Answer: **_c. Hot_**
Reference: Pages 6 and 10.
Explanation: The Hot leg is always the non-ground leg. It carries the current to the load. Normally that would also be the Positive leg. However, we define

Voltage potential as the difference of potential between two points, i.e. the Hot leg or positive of a Battery would be +12V while the negative or ground would be 0V.

What if you were dealing with a negative or -12V and 0V ground? In that case, the Hot leg would be the -12V and the more positive 0V would still be the ground.

On Positive Ground systems, the Hot leg would be the negative side of the circuit. Sound strange... Years ago, cars produced in Britain were all Positive Ground systems. They worked fine. The only problem came about when you tried to jump a positive ground car with a negative ground car. It could be done, but usually it was disastrous.

7. One of the most basic rules in electricity is that *Current* follows the path of least resistance.
Answer: *d. Current*
Reference: Page 11.
Explanation: Basic rule of electricity mentioned one way or another thru out this book.

If you and your body are more attractive (the path of least resistance) to the electricity then the previous ground path that it was using, then you will quickly become a <u>new part of the circuit and the new ground return path</u>. Refer to the illustration on next page.

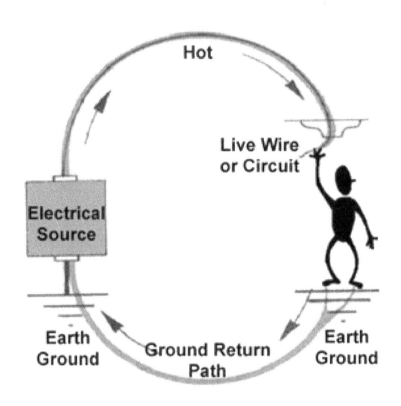

Hot

Live Wire
or Circuit

Electrical
Source

Earth
Ground

Ground Return
Path

Earth
Ground

8. "Normally open" means that there is no complete current path when the switch is in its ***Unswitched*** position. Answer: ***d. Unswitched***
Reference: Page 42.
Explanation: Switches are usually shown in wiring diagrams in their "normal" or unswitched position. Usually that will be the "OFF" position, but not always. "Normally open" refers to a switch (or relay contact) that does not complete or connect the circuit path. Normally closed identifies a switch that connects the circuit path in when it is in its unswitched position. In relays, the unswitched

position would mean when the relay coil is not powered.

9. The **_Primary_** side is the input side of a transformer. Answer: **_b._ _Primary_**
Reference: Pages 39 - 40.
Explanation: See image below.

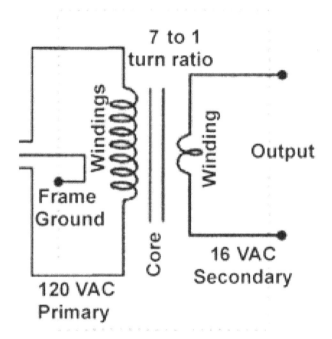

10. What do you call it when two points with different voltage levels (i.e. hot and ground) are connected with no resistance or load between the two points?
Answer: **_c. Short_**
Reference: Page 6.

11. Conductive articles of jewelry and clothing may be worn around exposed energized parts. Answer: *b. False*
Reference: Page 14.

12. A(n) *Open* will normally increase total circuit *resistance* and reduce total circuit current.
Answer: *b. Open; resistance*
Reference: Page 6.

13. A(n) *Short* causes more *current* and less resistance. Answer: *c. Short; current*
Reference: Page 6.

14. Which of the following is not something you should consider when forming a troubleshooting plan? Answer: *f. None of above*
Reference: Pages 57 - 60.
Explanation: All of these are things to consider when diagnosing the cause of a fault. You need all the help you can find. Consider every piece of info, no matter how trivial or seemingly unrelated.

15. Redundant circuits are second circuits that provide duplicate electrical paths. Why are they used?
Answer: *c. Increase battlefield survivability*
Reference: Page 5.

16. The influence or occurrence of "LRU Tolerances" Answer: *d. None of above*

210

Reference: Page 101.

17. What is a Mux Bus?
Answer: _**c. Different signals over one circuit path at same time**_
Reference: Page 115.

18. "Induced failure" relates to the theory that two inductive components (i.e. transformers) can pass a virus from one to another. Answer: _**b. False**_
Reference: Pages 104 –106.

19. The very first thing you should verify after turning an analog multimeter on.
Answer: _**e. Meter is Zeroed**_
Reference: Page 98.

20. You are checking continuity and resistance between two truck electrical connectors 21 ft apart. You decide to use the truck frame and chassis as the return line for your multimeter resistance reading. What may be a down-side or problem with using this technique?
Answer: _**c. Added resistance distorting reading**_
Reference: Page 24.

21. What is a better way of performing the continuity and resistance check in question #20.
Answer: _**c. Use external long wire for return**_
Reference: Page 24.

22. You have an <u>analog multimeter</u> to measure DC voltage across a device. You properly connect the probes to the terminals on the device, and the meter needle indicates 0 volts. This could indicate what?
Answer: *f. All of above*
Reference: Page 24.

23. You have a <u>digital multimeter</u> to measure DC voltage across a device. You properly connect the probes to the terminals on the device, and the meter indicates 0.00 volts. This could indicate what?
Answer: *f. All of above*
Reference: Page 24.

24. Which of the following is crucial to the correct operation of a Mux Bus?
Answer: *a. Unique component address*
Reference: Page 116.

25. If all other factors remain the same, what would be the effect of increasing from AWG (American Wire Gauge) 22 to AWG 26 in a wire run or circuit?
Answer: *b. Resistance will increase*
Reference: Page 80.
Explanation: The information you need to remember is that as the AWG increases the actual wire size decreases, i.e. AWG 10 wire has 4 times the current carrying area of an AWG 16 wire. This means that if you increase the AWG number of the wire in a circuit you also increase the resistance in that circuit. You will also increase the total resistance in a circuit if you considerably lengthen the wiring in the circuit.

26. Why do you shield wiring?

Answer: ***d. Reduce risk of outside EMF influence***

Reference: Pages 30 and 89.

Explanation: Unseen waves of EMI (electromagnetic induction) or EMF (electromagnetic field) are moving thru us and our environment all the time. Radio waves, cell phone signals, radar, lightning, just to name a few are all around. Normally they don't bother us. You have seen their effect on your TV when lightning strikes nearby. If the wave is extremely strong it can affect electronics by replacing the system signal with its own. You may have heard about a problem with some helicopters performing un-commanded movements when flying a few years back. This was finally traced down to EMI signals getting into the system. They went back and shielded the wires and the EMI problem went away. If the EMI is too strong the shielding protection can be overwhelmed. An example of this is the nuclear bomb EMP (electromagnetic pulse).

27. What type of circuit is shown below?

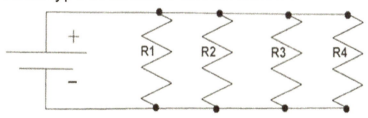

Answer: ***d. Parallel***

Reference: Page 5.

28. What type of circuit is shown below?

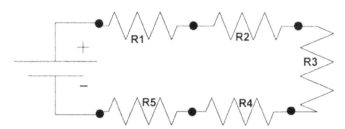

Answer: _c. Series_
Reference: Page 4.

29. In the following order, you close both CBs and move S2 to position 2. What occurs when you close S2, and why? (Diagram on next page).
Answer: _c. Top CB immediately opens, Short to ground_
Explanation: See diagram on next page. This circuit is has a positive ground instead of the more common negative ground. This arrangement does not make any difference. In fact, British cars were positive ground for many years with no issues (except when charging their battery). Closing one or both CBs and S1 with S2 in any position other than 3 creates a Hot to Ground short. This pops one of the two CBs depending on the 1,2 or 4 position of S2.

30. You again open both CBs and S2. In the following order, you close both CBs and move S2 to position 3. You close S2. What occurs and why?
Answer: _e. None of above_

Explanation: Absolutely nothing occurs because there is no connection to the battery Hot side. The circuit is one big loop that goes nowhere.

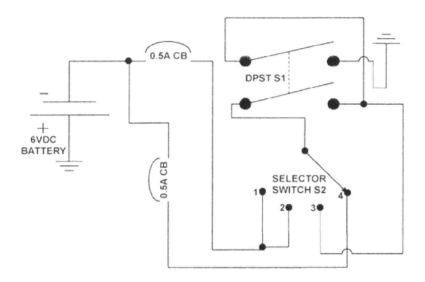

31. One method of troubleshooting involves using substitution or exchange. What are the benefits of using this method?
Answer: *c. Sometimes quicker*
Reference: Pages 69 - 74.

32. Low voltage systems have more intermittent connector problems than high voltage systems.
Answer: *a. True*
Reference: Page 79.

33. Which of the following is a function of a diode?
Answer: *e. Rectifier*

Reference: Page 83.

34. Which of the following are advantages of the Mux Bus? Answer: _**e. All of above**_
Reference: Page 115.

35. What is my suggested method of troubleshooting the Mux Bus?
Answer: _**c. Component exchange or substitution**_
Reference: Page 116.

36. Troubleshooting of wiring problems in a Mux Bus can be performed using a multimeter.
Answer: _**a. True**_
Reference: Page 116.

37. Using a multimeter, you check for continuity on the circuit path by placing the + red lead on P3303 pin M and the − black lead on J3303 pin P. The meter shows low ohms resistance. When you check for continuity by placing the − black lead on K1X1 and the + red lead on J3303 pin P you show what appears to be infinity. What is the fault?
Answer: _**a. There is no fault**_
Reference: Exercise wiring diagram (p.196-197, 203)
Explanation: The common component in both those circuit paths is the CR1 diode. A Diode is a one-way check valve. It allows or conducts electricity in one direction when forward biased, and blocks in the other direction when reverse biased. The forward bias direction of current is from the cathode thru the

anode (negative to positive). The anode is the arrow symbol. The simplest way to remember is that if the anode is positive, the diode was forward biased and will conduct. Refer to drawing below.

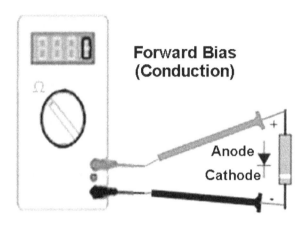

Forward Bias
(Conduction)

Anode
Cathode

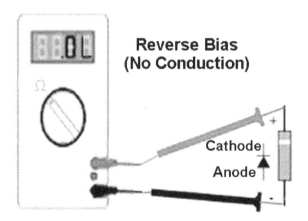

Reverse Bias
(No Conduction)

Cathode
Anode

In this example, when you placed the + red lead on P3303 pin M and the – black lead on J3303 pin P, you were forward biasing the CR1 diode and it conducted the multimeter current. This is indicated by the low ohms resistance reading. When you placed – black lead on K1X1 and the + red lead on J3303 pin P, you were reverse biasing the CR1 diode and minimal conduction occurred. This gave you the very high resistance reading. If there were no readings (infinity) or low resistance readings on both arrangements, the diode would be bad.

38. All CBs are open, TB3 pin 1 is now at 0 volts. K1 is in non-powered state. The light is off. S1 is in the normal position. Today is Monday. Friday will be the fifth day of the week. What is the resistance from TB3 pin 1 to chassis ground? Answer: ***d. Infinity***
Reference: Exercise wiring diagram (p.196-197, 203)
Explanation: This path should run from TB3 pin 1 thru <u>closed CB4</u> thru Relay K1X2 thru K1X1 thru TB2 term 2 to GN1. However, the question states "All CBs open" so the path ends at the input side of CB4. The correct answer is ***Infinity***.

39. CB4 is now closed. It is now Tuesday. The K1 coil now has 0 ohms resistance across K1X1 and K1X2. K1X2 has infinite resistance between it and TB2 term 2. What is the resistance from TB3 pin 1 to chassis ground? Answer: ***d. Infinity***
Reference: Exercise wiring diagram (p.196-197, 203)
Explanation: This path should run from TB3 pin 1 thru closed CB4 thru Relay K1X2 thru K1X1 thru TB2 term 2 to GN1. However, the question states "K1X2 has infinite resistance between it and TB2 term 2" so there must be a break in the circuit somewhere

between Relay K1X1 and TB2 term 2. The correct answer is *Infinity*.

40. TB3 pin 1 has +28vdc present. You close CB3. You close CB2. K1 coil resistance is 170 ohms. This is question 40. K1X1 is at ground potential. The resistances across K1B2 to K1B3 and K1A2 to K1A3 are both 0.8 ohms. You close CB4. The continuity reading across K1A2 to K1A1 is 0.8 ohms while K1A1 to K1A3 is infinite. All wiring is connected as shown on diagram. What is the resistance from J3303 pin A to P3303 pin H?

Answer: *d. Infinity*

Reference: Exercise wiring diagram (p.196-197, 203)
Explanation: You need to know four things on this one. 1-What is the status of Relay K1 when you take your final measurement? 2-Is it "OFF" with the contacts normal position (as shown on the diagram) or is Relay K1 "ON" with the contacts in the activated position (i.e. B1-B2 and A1-A2)? 3-How does J3303 pin A connect with Relay K1? 4-Is CB3 open or closed? How does P3303 pin H connect with Relay K1?

1- he Relay K1 has both +28V and ground, and 170 ohm coil so it will be in activated position when we do the resistance test from J3303 pin A to P3303 pin H.
2 & 3-J3303 pin A connects with Relay K1 thru closed CB3 at K1A2. 4) P3303 pin H connects with Relay K1 at K1A3. It all depends on what position contacts K1A1, K1A2, and K1A3 are in.... In this

case, the Relay K1 is activated and K1A1 and K1A2 are connecting. 4-Since there is <u>NO</u> connection between J3303 pin A and P3303 pin H, the correct answer is *Infinity*.

41. Refer to conditions set in Question #40. What is the voltage difference between the two CB4 terminals? What is the resistance between the two CB4 terminals? Answer: *c. 0 volts, none*
Reference: Exercise wiring diagram (p.196-197, 203)
Explanation: In this example, CB4 is closed and conducting. Since it is closed, both CB4 terminals would be at the same level of +28V with no voltage difference. The answer is *0 volts*. The only resistance between the two CB4 terminals would be that of the CB internals and would be about 0. The answer is *none*.

42. The ground potential at L1 terminals 1 and 3 is infinite. P3303 pin Z is now at +28vdc potential. S1 is in "ON". R1 has a 220-ohm value. L1 is a correct voltage serviceable correctly installed light. The resistance between L1 term 3 and TB1 term 1 is 100 ohms. What would it take to make L1 light up? (requires 2 answers)
Answers: *b. S1 in OFF position , d. Ground potential at J3303 pin c*
Reference: Exercise wiring diagram (p.196-197, 203)
Explanation: There are two issues here. The first is the +28V side. P3303 pin Z has +28V. That pin connects by wire to Switch S1 tem 2. When S1 is

"ON" the path would continue thru S1 term 3 to TB2 term 3 thru CB1 (if closed) out to J3303 pin f̱. This path never leads to Light L1.

However, if Switch S1 is "OFF", then the path would go from S1 term 2 across to S1 term 1 and continue thru TB2 term 4 to Light L1 term 1. So the first requirement and first answer is "**S1 in the OFF position**" so that +28V could reach Light L1 term 1.

For the ground side, the path goes from Light L1 term 3 to TB1 term 1 to J3303 pin c̱. The requirement for the ground side and second answer is "**Ground potential at J3303 pin c̱**".

The bulb has already been verified as good by the 100 ohms resistance reading between L1 term 3 and TB1 term 1. Add +28V and a ground and the light will be lit.

43. Assume that this is one part of a military aircraft. CBs 1, 2, and 3 are closed. The wiring from J3303 pin A to CB3 is completely cut apart by fragments from an enemy round. When you run a resistance check there still is continuity between J3303 pin A and K1A2. A second resistance check shows J3303 pin A is grounded. With the information provided along with the diagram as drawn, what is one possible cause of this?
Answer: _**c. Damaged wire is contacting the aircraft frame**_

Reference: Exercise wiring diagram (p.196-197, 203)
Explanation: This is case where you have to eliminate the wrong answers…. And you are left with the correct one. This is not "normal circuit operation" so answer "A" is out. Nothing shows that "K1A2 is grounded" so answer "B" is out. If K1A2 is not grounded then K1A3 is not either, so answer "d" is out. There is a possibility that the "Damaged wire is contacting the aircraft frame" and grounding the circuit, so "c" is the answer. By the way, "e" is out.

44. There is 28V at TB3 pin 1. Only CB4 is closed. You hear the relay contacts move. There is continuity between K1 terminals B2 and B3. There is also continuity between K1 terminals A1 and A2. You open CB4. There is 0.8 ohms resistance between K1 terminals A2 and A3, but infinity from A2 to A1. There is continuity between K1 terminals B2 and B3. Ground GN1 has failed and is now open. You again close CB4. There is infinity between K1 B1 and B2. You also close CB2 and CB3. What is the resistance reading from J3303 pin Z to P3303 pin F?
Answer: *e. 220 ohms*
Reference: Exercise wiring diagram (p.196-197, 203)
Explanation: There is +28V from TB3 pin 1 thru the closed CB4 to Relay K1. Relay K1 clicked when you closed CB4. This would give you a sign that it is working. You verify this with the continuity from the Relay A1 to A2 along with infinity from A3 to A2. If the Relay were not working, then the readings would

have been exactly opposite (continuity from A3 to A2 along with infinity from A1 to A2).

However, the readings from the B contacts do not match. There is continuity between K1 terminals B2 and B3, with infinity between K1 B1 and B2. This would indicate that the B1 and B2 contacts are shorted together and do not move when the Relay K1 coil power is On or Off.

So what do we now know? Relay K1 does power up, but only the A contacts move correctly. The Relay K1 B contacts are frozen or stuck with B2 and B3 always connected.

The conditions state that GN1 failed. Does that effect whether the Relay K1 operates or not? And if it does, will it change the answer to question #44? If GN1 were the only ground in this circuit, then its failure would shut down the Relay K1 coil. However, there is a second ground in this circuit connected to TB1 pin 3. Follow the wiring and you will see that they are connected together. The failure of GN1 will not effect Relay K1 coil operation (answer to question #45). If the second ground fails, then Relay K1 coil would not power-up. But does any of this effect the answer to question #44?

The question asked "What is the resistance reading from J3303 pin Z to P3303 pin F?" The correct path is as follows: J3303 pin Z to wire split (1 wire path is redundant wiring for back-up), both paths provide a path to closed CB2. The path continues to K1 B2. if the Relay K1 is operating, and the B contacts move as designed, then the circuit stops at the K1 B1 contact. However, from the previous information, you know that the K1 B contacts are stuck with contact B2 always connecting to B3.

This allows the path to continue from K1 B2 across to K1 B3 up to TB1 term 4 thru Resistor R1 to TB1 term 5 to P3303 pin F. The answer to question #44 is the value of R1 or 220 ohms. The answer to the question "whether Relay K1 having power and ground to function will change the answer to question #44" is no it will not.

The answer to the original question is determined by the connections of the Relay K1 B contacts. In this case since the B contacts are stuck and fixed, that determines the correct answer. However, if the B contacts were functioning as designed, then the correct operation of the Relay K1 coil would be pertinent.

45. Same conditions as Question #44. What effect does the failure of Ground GN1 have on wire shielding? Answer:_ *c. No effect*
Reference: Exercise wiring diagram (p.196-197, 203)
Explanation: See question #44 explanation.

46. You are checking for continuity. You find 150 ohms resistance between J3303 pins c and f. Switch S1 is in the "ON" position. There is continuity between P3303 pin A and J3303 pin D. There is continuity between P3303 pin Z and R, but infinity relative to any other pins in either P3303 or J3303. What is the most probable fault?

Answer: *c. Switch S1 burned out*

Reference: Exercise wiring diagram (p.196-197, 203)

Explanation: If you trace the paths from J3303 pins c and f, you end up at S1 contacts 1 and 3. These two contacts are never supposed to connect. That is a fault. If you trace the circuit path from P3303 pin A and J3303 pin D, it does connect thru S1, but only when S1 is "OFF." However, the question told us that S1 is "ON." That is another fault. P3303 pin Z and R (basically the same wire) have no continuity to any other wires, but they should connect up with either J3303 pins c and f depending on the S1 position. That is one more fault. All 3 faults indicate a burned out S1 switch with contacts 1 and 3 shorted together, the wiper arm at 2 melted away, and a stuck 5 – 6 – 4 wiper arm.

47. Refer to Question #46. Would your answer be different if Switch S1 was in the "Off" position?

Answer: *b. No*

Reference: Exercise wiring diagram (p.196-197, 203)

Explanation: Switch S1 is burned out. It makes no difference what position it is in.

Questions #48, #49, and #50 use the following conditions. (Careful... this is tricky).

You know, observe, or have measured the following:
- K1X1 to chassis ground resistance 170.7 ohms
- +28vdc at TB3 pin 1
- CB3 and CB4 closed
- CB1 and CB2 are open
- +28vdc at K1X2
- K1B3 to K1B2 resistance infinite
- Continuity from K1B1 to K1B2
- Infinite resistance K1B3 to K1B1
- Resistance K1A3 to K1A2 approx 0.7 ohms
- CR1 is open and faulty
- +28vdc at J3303 pin A
- P3303 pin R and P3303 pin H are electrically jumpered together (connecting wires)
- S1 is in ON position
- P3303 pin Z and P3303 pin W are electrically jumpered together (connecting wires)
- J3303 pin K and J3303 pin f are electrically jumpered together (connecting wires)
- TB2 pins 3 and 4 are electrically jumpered together (connecting wires)
- TB1 pins 1 and 3 are electrically jumpered together (connecting wires)
- L1 is lighted

48. What effect does the dimmer switch have on L1?
Answer: *c. No effect*

Reference: Exercise wiring diagram (p.196-197, 203) Explanation: Since there is 28VDC at K1X2 with continuity of 170 ohms thru the coil and a ground, K1 relay should be operational. This can be verified by checking the continuity between the contacts of A1-A2-A3, and B1-B2-B3. There should be continuity from A1 to A2 and from B1 to B2. Measurement "g" verifies contacts resistance from B1 to B2 as continuity or 0 ohms while Measurement "f" gives us infinity from B3 to B2 and Measurement "h" indicates infinity from B3 to B1 as expected.

However, Measurement "i" gives us 0.7 ohms resistance from A3 to A2 when there should be continuity from A1 to A2. Since the B1-B2-B3 contacts did change with the operation of relay K1 coil movement, the incorrect reading from A3 to A2 tells us that those contacts are bad and shorted or stuck together (first answer to question #50, CR1 open (as stated in Measurement "j") is the second answer to question #50). The primary path for +28VDC getting to Light L1 is as follows: J3303 pin A at +28V to split in wire (1 wire path is redundant wiring for back-up), both paths provide power to CB3, which is closed.

Power continues to Relay K1A2 contact (which is stuck) to K1A3 contact. +28V continues to split in wire (1 wire path is redundant wiring for back-up), both paths provide power to P3303 pin H, which is jumpered to P3303 pin R. Power continues to Switch

S1 (which is in the on position) term 2 and continues across to term 3. +28V continues to TB2 term 3, which is jumpered to TB2 term 4. Term 4 is directly wired to Light L1 term 1. L1 gets its required ground thru L1 term 3 thru TB1 term 1 (which is jumpered to TB1 term 3). TB1 term 3 is a grounded terminal. That is the primary path to Light L1.

However, what happens if you change Switch S1 to the "OFF" position. The +28V path to Switch S1 term 2 remains the same as above. If S1 is off, then S1 term 2 connects with S1 term 1. Power would continue on the wire to TB2 term 4, which is directly wired Light L1 term1. The ground circuit is the same. What does this show? It shows that in this case, both positions (ON or Off) provide +28V to Light L1. There is no difference if Switch S1 is ON or OFF (answer to question #49).

There is a dimmer circuit (question #48) also indicated and it is as follows: The +28V path still goes to Switch S1 term 2 and remains the same as above. However, there are 2 wires connecting to Switch S1 term 2. Power also moves to P3303 pin Z, which is jumpered to P3303 pin W. +28V, moves thru TB1 term 2 the Dimmer term 1. The Dimmer varies the voltage going out Dimmer term 2. This variable voltage continues to J3303 pin K that is jumpered to J3303 pin f. The variable power continues on until it reaches CB1. CB1 is open so that is the end, and the dimmer has no effect on Light L1 (answer to #48).

However, what if CB1 were closed, would it have an effect on L1 then? From the earlier explanation, you know that there is +28V at TB2 term 3 at all times. CB1 is directly wired to TB2 term 3, so if CB1 were closed, you would have a +28V meeting a variable voltage less then a +28V….with absolutely no effect on Light L1. This answer is still no effect.

49. What effect does the on or off position of Switch S1 have on L1? Answer: ***b. No effect***
Reference: Exercise wiring diagram (p.196-197, 203)
Explanation: See question #48 explanation.

50. What components have failed? (Two answers)
Answers: ***c. K1A2 & A3 are shorted or stuck together***, ***d. CR1 open***
Reference: Exercise wiring diagram (p.196-197, 203)
Explanation: See question #48 explanation.

How did you do? Make sure that you go back and read the explanations if you have a different answer.

To discuss how I arrived a particular answer, email me at:

Dar-Bet@att.net

For more about troubleshooting, visit my website at

BasicTroubleshooting.com

To learn about more of my books, see the final four pages of this book or visit my website at

DarrelKaiserBooks.com

Summary

I hope this book answered most of your questions and helped clear up the mystery of troubleshooting. Troubleshooting is a skill that you only improve with practice. This is one of the problems with depending on ready-made fault Trees. You do not have to really think, because you are only following someone else's instructions. With a ready-made fault tree, your troubleshooting success <u>is always dependent on their troubleshooting skill</u>.

A wise man once wrote, "Practice your own troubleshooting and improve your skills for one day there may not be a Fault Tree available." That same wise man also wrote "Remember, don't assume anything, verify everything, and think out of the box."

And last, the wise man wrote, "Do not ever allow yourself to become the ground return path."

Good Luck!

Darrel P. Kaiser

BasicTroubleshooting.com
DarrelKaiserBooks.com

Books by Darrel P. Kaiser
www.DarrelKaiserBooks.com

Origin & Ancestors
Families
Karle & Kaiser
of the
German-Russian Volga Colonies

Adolf	Heytmann	Ruder
Andreas	Hieronymus	Rudolph
Arp	Horn	Schaeffer
Arnst	Ikstadt	Schörer
Becker	Kaiser	Schiller
Bopp	Kurie	Schmiede
Barbach	Köhler	Schneider
Dayenheim	Krämer	Schütz
Folit	Lieders	Simon
Freund	Maurer	Sielz
Geringer	Michel	Trieber
Grün	Neff	Trippel
Hart	Neumann	Vogt
Heiland	Nicolausen	Werner
Hermans	Nilmayer	Will
Hess	Popp	Zeichmann

Darrel Philip Kaiser

**Moscow's
Final Solution:
The Genocide
of the
German-Russian
Volga Colonies**

Darrel Philip Kaiser

**Religions
of Germany
and the
German-Russian
Volga Colonies**

Darrel Philip Kaiser

**The Bad
and
Downright Ugly
of the
German-Russian
Volga Colonies**

Darrel Philip Kaiser

Books by Darrel P. Kaiser
www.DarrelKaiserBooks.com

Emigration to and from the German-Russian Volga Colonies

Darrel Philip Kaiser

Basic Electrical Troubleshooting for Everyone

Darrel Philip Kaiser

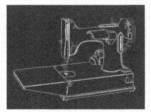

the Featherweight

Ads

Darrel P. Kaiser

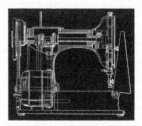

the Featherweight

Patents

Darrel P. Kaiser

Books by Darrel P. Kaiser
www.DarrelKaiserBooks.com

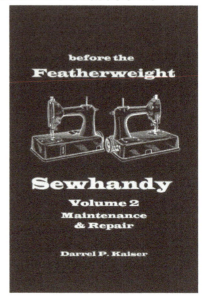

Logical
Sewing Machine
Troubleshooting

for Everyone

Darrel Philip Kaiser

Books by Darrel P. Kaiser
www.DarrelKaiserBooks.com
Coming in 2008…

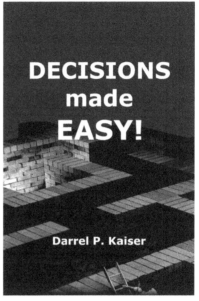

For more information, visit my websites at:

DarrelKaiserBooks.com
Volga-Germans.com
SewingMachineTech.com
BasicTroubleshooting.com
SewhandySewingMachine.com

www.ingramcontent.com/pod-product-compliance
Lightning Source LLC
Chambersburg PA
CBHW051230050326
40689CB00007B/866